The Power of Water

A Primer for Anyone Entering the Water Industry

PETER STYLES

AuthorHouse™ UK
1663 Liberty Drive
Bloomington, IN 47403 USA
www.authorhouse.co.uk
UK TFN: 0800 0148641 (Toll Free inside the UK)
UK Local: 02036 956322 (+44 20 3695 6322 from outside the UK)

This book is printed on acid-free paper.

ISBN: 978-1-7283-7461-1 (sc)
ISBN: 978-1-7283-7462-8 (e)

Print information available on the last page.

Published by AuthorHouse 08/31/2022

authorHOUSE®

Special thanks to Alan Sutton for his major contributions especially on Distribution and Materials and Wayne Earp for his valuable insights

All unaccredited photographs are by Peter Styles.

All accredited photographs are via Creative Commons, to whom a donation has been made. The links to their sources can be found at:

https://felixschrodinger.wordpress.com/2022/07/21/the-power-of-water/

Thanks again to Wikipedia, Wikimedia, Creative Commons and all of the free sites who make knowledge sharing so easy these days.

An apology is due for the poor quality of some of my own images. Two of the major water companies were approached for help but both declined.

Dedicated to all who enjoy sharing knowledge and, especially, those who do it for free.

Contents

Also by Peter Styles (writing as Stilovsky and Schrodinger)

Hoggrills End **published December 2017**

The Power of Numbers **published January 2018**

The Power of Names **published May 2018**

The Power of Notes **published September 2018**

The Power of Words **(1) published December 2018**

Power Quiz '18 **published January 2019**

Power Quiz '19 **published March 2019**

Power Quiz '17 **published July 2019**

The Power of Words (2) **published March 2020**

The Power of Words (3) **published May 2020**

The Power of Dreams **published July 2020**

SAMS, Simplified Asset Management Systems **published December 2020**

Principles of Asset Management **published 2021 with Wayne Earp**

[all available in paperback or Kindle on Amazon]

Foreword

Working on New Civil Engineer's 50th anniversary issue this year has shown me that skills shortages were a concern in 1972, the same as they are now, and it is likely to be the same in another 50 years, which is why sharing knowledge with those who have joined the sector is so critical. This is the key reason why books like *The Power of Water* are essential building blocks for our industry – both now and in years to come. It draws together the many years of learning from retired engineers and stores it for future generations to ensure that lessons learned are not lost.

This need to continue learning beyond your formal education is very pertinent for me this year as it marks 25 years since I graduated and went into my first job as a geotechnical engineer. I had the benefit of studying at a former polytechnic that maintained its focus on delivering practical learning despite the transition to university status. Nonetheless, I can still remember hanging up my graduation gown and donning my steel toe-capped boots and hard hat with the realisation of how much I still had to learn. And I have never stopped learning since, even with my move to construction journalism.

What I do remember clearly from my time as a new graduate in 1997 is the time some of the senior staff took to explain processes, industry history and basic business essentials to me. I then had the chance to build on that knowledge and become more self-reliant working as a resident engineer. Sadly, the economic crisis of 2008 and 2009 changed the civil engineering industry forever with many graduates expected to be earning from day one and the role of resident engineer mostly having disappeared.

The loss of these learning opportunities makes books like *The Power of Water* even more essential reading as the chance to gain such insight on the job is becoming rarer. Peter recalls "sitting with Nellie" during his formative years and I recall "sitting with John" in the same way but I hope that for you, "sitting with *The Power of Water*" will create a firm foundation for your career in the water industry.

As an industry we have come a long way since Sir Joseph Bazalgette transformed the health of Londoners with his sewer system and we are about to see another step change with the commissioning of the Tideway Tunnel. However, there is much to learn about the drivers for that change and how the industry has developed. It continues evolving for the better and having skilled employees is a key part of that evolution.

Claire Smith
Editor
New Civil Engineer

Introduction

Why write a book about water?

Having time to spare, I set about looking at what's available to a new entrant to the industry to cut their teeth on. And the answer is very little - yes, there are a number of excellent courses and a few informative books, but there's nothing that covers the whole subject at the basic level, and is readily available at minimal cost. So, my answer to the above question is – 'knowledge sharing' without you having to break-the-bank to access it.

I remember well, my days "sitting with Nellie" as I was learning about the basics of the sewer system in Manchester and would have been so grateful for a book which would provide much of what "she" was showing me. How little I knew in those days – and how much were things to change over the coming years.

A short introductory, on-line, course could cost you hundreds of pounds and most specialist textbooks can cost as much. Searching out papers on the internet may be cheap but it's very time consuming. So, I have aimed, through self-publishing, to produce a broad-based textbook at a fraction of those costs and almost free if you use the downloadable Kindle version. My concentration is on getting a complete broad-based coverage rather than an in-depth explanation of any of the subjects.

Having been involved in safety training and the development of the BTEC courses in my days at the Leicester Water Centre, I found that I have much to share. Whilst this book is aimed at new entrants, I hope that it will provide a valuable insight to others who may already be employed in associated non-technical fields such as finance and human resources. In addition, it should help those in developing countries to appreciate the basics of water, its technologies and its management.

Whilst I have concentrated on the UK Water Industry, most of the content is relevant elsewhere and of interest to anyone who wishes to learn about water and how mankind makes use of it. I apologise for not writing too much about Scotland, Wales and Northern Ireland but most of what's in here is common to them as well.

You may wish, before embarking on reading my work in its entirety, to find, in the contents list, where the glossary and list of abbreviations are so that you may refer to them when necessary. And don't expect lots of stuff on the latest gadgets as the author is long past his sell-by date.

At an entry level, basic skills and organisation are crucial and this is where I have aimed this book. I call it a 'primer' but expect it to be useful to anyone who is new to the industry or just wants to know more about how things work. It is based on many years of experience in local government and the water industry including at home and abroad.

I hope you find it useful and good luck in your career.

Background

The water industry in the UK is complex and has undergone many changes throughout its history. These involve the organisation and management of its functions alongside the technological advancement which have taken place, in which the UK has played a pivotal role. Whilst this book concentrates on the UK industry, most of its content is relevant throughout the world if only in contrast to what happens elsewhere.

What do we mean by 'the water industry?' Well, most of you will probably think of the companies that supply you with drinking water and take away your sewage. Yes, that's right to a large extent, but there is a wider context and I have tried to encompass all, or most, of the other players who are responsible for water matters.

The roots of the industry are founded in local government and we remain grateful for the foresight of the great city fathers, especially of the Victorian era for much of the infrastructure that we rely on today. In referring to this heritage, we must be mindful that the wealth created by the industrial revolution made much of this possible though many would emphasise other aspects of our country's history. Virtually all of the utility industries – water, sewerage, gas, electricity, local transport and even telecoms have their roots in the local councils who ran their area. It is only since the latter part of the 20th century that they became public companies and hence responsible to shareholders.

Whilst it was a matter of local judgement as to what services should be provided, it's a relatively recent thing for services to be prescribed in law or regulation. The 1945 Water Act and the 1936 Public Health Act were pivotal in moving things forward until overtaken by the 1973 Water Act which set up the framework, much as we now see it. When lack of government finance was perceived as a major problem in the 1990s, privatisation brought a new dimension to the industry and it remains largely in that form today.

Much can be learned from contrasting water rights in the UK with those in the USA. All water abstraction in the UK is controlled by a delegated government department using regulations based on legislation and, with few exceptions, everyone knows where they stand. In contrast, the American system is still based on historical 'water rights' which tend to give absolute power to the landowners. In India, states regularly go to the High Court to resolve water disputes.

Like most public services, water depends on the quality of its governance. In this respect, the quality of management, the availability of technical skills and the independence of those structures from financial and political in-fighting is crucial. There remains a fine line between the benefit of the majority versus the rights of the individual and this is ever present in our industry.

The Macro and Micro Water Cycles

The Blue Planet

Earth is sometimes referred to as 'The Blue Planet due to its appearance from space. This is due to a large proportion of its surface being covered by liquid water.

The Macro Water Cycle

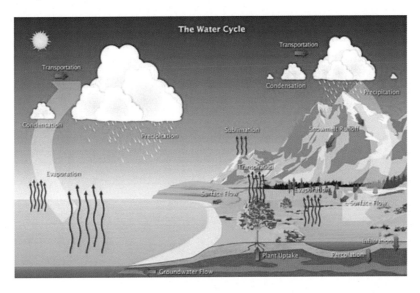

Atmospheric Infrared Sounder

The 'macro cycle is the natural way that rain falls onto the land, flows to the sea via rivers and streams and then evaporates to form clouds which, in turn, create rainfall.

The Micro Water Cycle

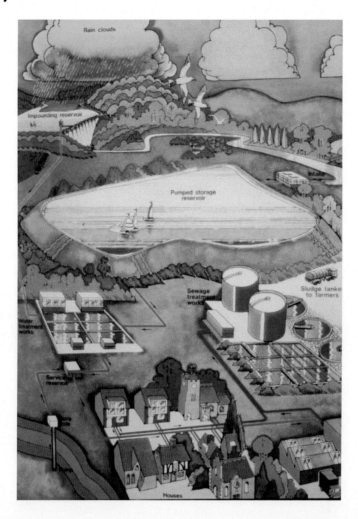

The 'micro' water cycle concerns mankind's use of the water after it falls as rain and then, eventually, gets discharged back to the environment. After it has fallen it enters the ground (ground water) or streams which become rivers. It is stored and abstracted before treatment making it 'potable' – i.e. suitable for drinking (from the Latin 'to drink'). After use, it is collected and transported to treatment before it is discharged back to the environment, usually into rivers.

Whilst all of this looks simple in outline, it requires established management making use of extensive installations and technical manpower as well as energy to get it to where it is needed. These are the essentials that make the difference between systems that work and those that don't.

A History of Water in the UK

Whilst elsewhere in this book, I have looked at the wider context of the water industry, here I have generally restricted myself to the narrower aspect of water supply and sewerage.

Long before man became man (and woman), the essential nature of water was appreciated even in the animal world. Elephants have trumpeted its value for aeons and Hippos likewise – but ask yourself – have you ever seen a rhinoceros take a bath? Having descended from a wolf, your dog will suffer you bathing him but that's only because you feed him. Your cat will pass and tell you that he's already had a good lick that morning and deposited his bodily waste in the neighbour's flower bed. But why is it that your dog can happily drink polluted water and even consume the droppings of other species without getting ill whilst we get sick at the very thought of it? Obviously we have very divergent evolutionary paths but why are you wasting so much money on expensive dog food?

Jean M Auel, in her book *The Clan of the Cave Bear*, and the ensuing *Earth's Children* series, described much of mankind's development and many of his/hers primitive inventions. Despite popular fiction, it's not clear when it became apparent that polluted water was a common cause of disease but, even before we had writing, mankind was aware that clean water was essential to a healthy life. They took their water from upstream and deposited waste downstream.

All flora and fauna require water and by instinct animals, including humans, have established their homes and settlements within easy reach of a water supply. The importance of constructing reliable, clean water supplies were understood as far back as 8,000 BCE although it's likely that, even earlier, man understood that tasteless, cool, odourless and colourless water was considered the healthiest, and that stagnant, marshland water was to be avoided.

Initially early man established settlements near lakes, rivers or streams in order to have easy access to water. Later, he realised that sometimes water can be found below ground even if none is showing on the surface and in the Jezreel Valley in Israeli wells have been discovered which date as far back as 6500 BCE. Water would be drawn and then carried by hand to be used for cooking and washing. Evidence exists in Peru of an interconnected well and underground watercourse dating back to about 3200 BCE.

The ancient Greeks were possibly the first civilisation to use clay pipes for water transmission, approximately 2000 BCE. However, it is the Romans to whom we generally give thanks, especially in the UK, for their amazing engineering capabilities and foresight in constructing tremendous aqueducts from stone together with pipelines using hollowed-out tree trunks (hence the term 'trunk main'), usually elm, for water and wastewater transmission. Not content with simply supplying water for cooking and bathing the Romans also developed water features and fountains for citizens to enjoy. They also developed indoor plumbing using lead pipes, which were still used in many countries, including the UK, until the 1960s. The Romans, and the ancient Greeks, often improved the water quality by the use of settling tanks, sieves, filters and they boiled water.

There are two very notable moments in history - both attributed to Archimedes. Everyone was aware that moving water uphill was problematic but he was someone who sought a solution. He is credited for inventing the screw pump though he may actually have made an improvement on an earlier design as they were said to have used pumps to water *The Hanging Gardens of Babylon*. How to follow that for an idea? Well few have genuine 'eureka' moments but he did when he thought about how to measure the volume of an irregular solid – immerse it in water (obvious when you think about it!). Thus was created the principle of buoyancy.

Whenever you go to visit an ancient monument building, you may be confronted with details of their water supply and sewerage arrangements. Almost every bit of writing on this subject mentions the Romans and their aqueducts, even something about their waste disposal systems. This often leads us to believe that the Britons, under Roman rule, had access to potable water and drains – they didn't. Traces of these water systems from ancient times survive because they were well built to serve those in power. Other than those who served the masters most would probably not even have known that these systems existed let alone been able to make use of them.

Lead pipe by Gds

So let's keep things in context. Primitive society drew its water from any available watercourse until it was found that using a well to draw groundwater from the local aquifer provided some protection against disease. The amount of water used was so small that sophisticated disposal systems were not needed. Defecation would normally have involved a midden and urination would have been virtually anywhere.

After the Romans left Britain, the road network and their water structures fell into decay. The Britons knew nothing about asset management and just went back to the land. Whilst the Anglo-Saxons had many aspects of culture, waterworks do not seem to have featured strongly amongst them. So we have to get to the Normans before building once more becomes a pastime for the nation. But, again, we see that the landed gentry have access to facilities but the general population are left to get on with it. They didn't even speak the same language. Norman castles were generally constructed with a defensive moat which also served as an open cesspit. At many castles, Kenilworth for instance, you can still see the remains of the toilet which was built

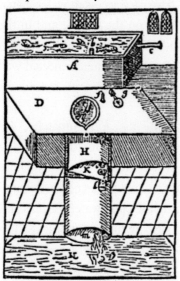

A priuie in perfection

A. the Cesterne.
B. the little washer.
C. the wast pipe.
D. the seate boord.
E. the pipe that comes from the Cesterne.
F. the Screw.
G. the Scallop shell to couer it when it is shut downe.
H. the stoole pot.
I. the stopple.
K. the current.
L. the sluce.

M.N. the vault into which it falles: alwayes remember that ( ) at noone and at night, emptie it, and leaue it halfe a foote deepe in fayre water. And this being well done, and orderly kept, your worst priuie may be as sweet as your best chamber. But to conclude all this in a few wordes, it is but a standing close stoole easilie emptied.

And by the like reason (other formes and proportions obserued) all other places of your house may be kept sweet.

into the internal face of the external wall. One of the reasons why the court moved on from castle to castle in medieval times was that, besides expense, the moat would stink to high heaven due to the excess load of faeces which it could not treat.

Even in medieval times, it was appreciated that drinking water was risky and the populace resorted to relying on weak beer as the process of boiling the water before brewing made it much safer. The later addition of hops made it last longer and so relieved many a housewife from the daily chore of brewing.

Some credit Elizabeth I's godson John Harington with inventing the flush toilet but it was hardly that – simply a boxed seat to sit on and defecate, after which a bucket of water was used to flush the contents away. The call of "Gardez l'eau" was prevalent in many urban areas as someone tossed their wastewater out into the street.

Wooden pipes were made from tree trunks, mainly elm, bored through from end to end. A project to bring fresh water from Hertfordshire to London by Sir Hugh Myddleton brought about London's first public water supply system in the construction of the "New River" between 1609 and 1613. The New River Company became one of the largest private water companies of the time, supplying the City of London and other central areas and it continues to supply River Lea water to Londoners today. The first civic system of piped potable water was from the River Derwent to Derby in about 1692 using wooden pipes like the one illustrated (Wikimedia).

In 1778, inventor Joseph Bramah patented a flush toilet one of which can still be seen working at Osborne House. It's said, that his name provided the origin for our use of the word 'Bramah' as being something very special. If you don't know of it - just ask a bridge player.

Sand Filtration and Public Supply

The first sand filters for purifying the water supply date back to 1804, when John Gibb a factory owner in Paisley, installed one to recycle his used water and sold it to the public. This was refined in the following years by private water companies and was then used by The Chelsea Waterworks Company in London – who provided filtered water in 1830 for residents of the area.

Whilst there was much going on to improve the transport of goods around the country and abroad, where was the massive investment in the needs of the populace for clean water? Obviously there are isolated examples of philanthropy such as New Lanark (1786) and Saltaire (1851) but the majority of poor people continued to be in want of reliable supplies. Bourneville was to come much later.

Crapper

Then along came Thomas Crapper who is credited with all sorts of things including the phrase "Going for a crap." This is reputedly based on "I'm going to visit Mr Crapper" when the need took you as his name was emblazoned on the toilet bowl. Whilst it's not clear if anyone actually 'invented' the flush toilet, it's quite certain that he 'developed' it to the point where it actually worked. Prior to his work there was excessive wastage due to flush toilets being left to run continuously. The Water Board looked out for a solution and Crapper produced his 'Valveless Water Waste Preventer' which incorporated a cistern which shut off the supply when full and we still use something similar today.

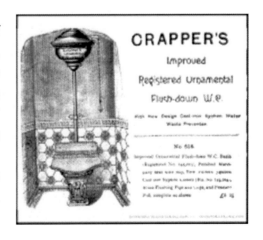

Sybil; or The Two Nations

I have covered some of the advances that were made in the supply of clean water and the collection and treatment of human waste. What this does, however, is to paint a false picture of what life was like for the ordinary folk who could no more afford one of Crapper's toilets than my grandparents could own a car – or even their own house. Benjamin Disraeli published his novel in 1845 – the early days of Victoria's reign as he aspired to address the sqaulid conditions that the ordinary people endured. This is an extrct, courtesy of The Gutenberg Project, from his book.

The situation of the rural town of Marney was one of the most delightful easily to be imagined. In a spreading dale, contiguous to the margin of a clear and lively stream, surrounded by meadows and gardens, and backed by lofty hills, undulating and richly wooded, the traveller on the opposite heights of the dale would often stop to admire the merry prospect, that recalled to him the traditional epithet of his country.

Beautiful illusion! For behind that laughing landscape, penury and disease fed upon the vitals of a miserable population!

The contrast between the interior of the town and its external aspect, was as striking as it was full of pain. With the exception of the dull high street, which had the usual characteristics of a small agricultural market town, some sombre mansions, a dingy inn, and a petty bourse, Marney mainly consisted of a variety of narrow and crowded lanes formed by cottages built of rubble, or unhewn stones without cement, and from age, or badness of the material, looking as if they could scarcely hold together. The gaping chinks admitted every blast; the leaning chimneys had lost half their original height; the rotten rafters were evidently misplaced; while in many instances the thatch, yawning in some parts to admit the wind and wet, and in all utterly unfit for its original purpose of giving protection from the weather, looked more like the top of a dunghill than a cottage. Before the doors of these dwellings, and often surrounding them, ran open drains full of animal and vegetable refuse, decomposing into disease, or sometimes in their imperfect course filling foul pits or spreading into stagnant pools, while a concentrated solution of every species of dissolving filth was allowed to soak through and thoroughly impregnate the walls and ground adjoining.

These wretched tenements seldom consisted of more than two rooms, in one of which the whole family, however numerous, were obliged to sleep, without distinction of age, or sex, or suffering. With the water streaming down the walls, the light distinguished through the roof, with no hearth even in winter, the virtuous mother in the sacred pangs of childbirth, gives forth another victim to our thoughtless civilization; surrounded by three generations whose inevitable presence is more painful than her sufferings in that hour of travail; while the father of her coming child, in another corner of the sordid chamber, lies stricken by that typhus which his contaminating dwelling has breathed into his veins, and for whose next prey is perhaps destined, his new-born child. These swarming walls had neither windows nor doors sufficient to keep out the weather, or admit the sun or supply the means of ventilation; the humid and putrid roof of thatch exhaling malaria like all other decaying vegetable matter. The dwelling rooms were neither boarded nor paved; and whether it were that some were situate in low and damp places, occasionally flooded by the river, and usually much below the level of the road; or that the springs, as was often the case, would burst through the mud floor; the ground was at no time better than so much clay, while sometimes you might see little channels cut from the centre under the doorways to carry off the water, the door itself removed from its hinges: a resting place for infancy in its deluged home. These hovels were in many instances not provided with the commonest conveniences of the rudest police; contiguous to every door might be observed the dung-heap on which every kind of filth was accumulated, for the purpose of being disposed of for manure, so that, when the poor man opened his narrow habitation in the hope of refreshing it with the breeze of summer, he was met with a mixture of gases from reeking dunghills.

Snow and Chadwick

After a cholera outbreak in Broad Street, London, in 1854, John Snow mapped the locations of the sufferers. He identified that the main cause as the local pump which was supplying infected water as it had been installed only feet away from an old cesspit. After he had the pump handle removed, the epidemic ceased. It sparked a call for better supplies of drinking water and means of sewage disposal. Local brewers – who had free beer – were untouched by the epidemic.

Replica of the Broad Street pump

Lancastrian, James Chadwick moved to London and become involved in public health. In 1842 he produced his: *Report on the Sanitary Condition of the Labouring Poor.* Whilst many opposed his reforms as Commissioner, he became responsible for setting the standards which form the bedrock of our sanitary systems today. The regulations of 1855 demanded that water must be of a certain quality and the first use of chlorine followed in 1897 to keep the public water supply pure.

By the end of the 19th century, piped-in treated water made drinking from public pumps and fountains safe for the first time in England.

In these times there were a number of strange constructions. As the supply of water was low, sewers needed to be flushed and 'flushing chambers' were built at the head of systems. The 'tippler' was a type of toilet used in the north which had a tray which retained the contents (faeces and urine) until it tipped over when full. This was reputed to allow the sewage to flow more freely but some would disagree. Many residents feared that the odours from foul sewers would harm them and 'disconnection chambers' were built to isolate their drain from the sewer.

Bazalgette

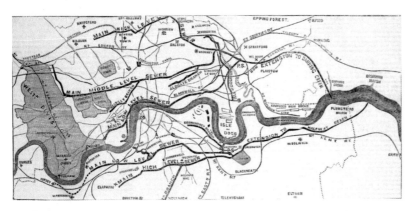

The main advantage of the toilet was that it took the unpleasant results of defecation away without the smell but it had less fortunate side effects. It required a considerable amount of clean water to function so the demand on the public supply was increased substantially. Also, there was the need to connect the outflow to another pipe to carry it away – thus the need for drains and sewers. In the early days the outflow from many toilets was not taken to treatment and ended up in the nearest watercourse. The "Great Stink" in the summer of 1858 led to Parliament being threatened with closure as it was so bad and even the Queen cut a trip on the river short because of it. On the plus side, it led to Joseph Bazalgette being commissioned to solve the problem. After years of delay due to the expense, he built the interceptor sewers which lie beneath the London Embankments. The intercepted sewage was taken to a pump station along the north bank to Abbey Mills and on to a treatment plant at Beckton. The south bank sent its sewage for treatment at Crossness and both plants still function today.

When Queen Victoria visited Trinity College in Cambridge, she asked what were all the pieces of paper floating down the Cam. Her guide remarked "Those, ma'am, are notices that bathing is forbidden." At that time the Fleet River was an open sewer and was not culverted until 1844. In the same year, Windsor Castle was found to have a multitude of overflowing cesspits. Typhoid and Smallpox were rampant at the time and cholera was to follow. Prince Albert contracted typhoid and died in 1861.

Thomas Twyford is credited with bringing about a major improvement in the design of the flush toilet – the 'combination' which led to many improvements and even, later, the close-coupled unit. Whilst the new toilets improved things – at least for those who could afford them - there was a big problem with smells coming back from the sewer. To solve this many installed a 'disconnecting trap' between their drain and the sewer to prevent odours coming back up. Later, Crapper was also involved in the development of urinals and in 1891, patented a toilet which flushed when you lifted the seat.

When his niece was asked by biographer, Wallace Reyburn, whether she was ever embarrassed by bearing his name, she responded "Good heavens, no – only the other day I was in Westminster Abbey and there beneath my feet was his name." "In Poets' Corner?" Reyburn asked. "No," she responded, "On a manhole – he did the drains for the Abbey, you know."

Night Soil

Whilst we write about the development of the water-born systems, we ignore the common system in poor areas of cities which endured right into the 20th century. 'Back-to-backs' were not Coronation Street houses but those that backed on to another house and hence had no back door; they usually had an address like 'No 5, Back Smith Street and frequently formed a court with a hand pump at the centre. Some examples survive, care of The National Trust, in Hurst Street, Birmingham.

These properties generally had no access to a sewer because they produced little liquid waste and hence relied on the municipality to collect their 'night soil' which was a mixture of faeces, urine and ash from the fire. The proceeds were collected, generally on horse drawn carts and taken to the outskirts of the city where they were simply tipped on to land and left to decay naturally.

Case Study – Nottingham Water Supply

The first water supply company was the Nottingham Waterworks Company, established in 1696, which took water from the River Leen, and later from springs at Scotholme, when the river became polluted. Other companies were established over the years and, in 1826, The Trent Water Company was formed with Thomas Hawksley as engineer. The various companies amalgamated in 1845.

Acquisition of the water company by Nottingham Corporation was first considered in 1852, but the water company initially resisted the proposals; the takeover eventually happened in 1880, and the Nottingham Corporation Water Department was created. Nottingham became a city in 1897, and the water department was renamed as the City of Nottingham Water Department in 1912. The Corporation co-operated with Derby, Leicester, Sheffield and Derbyshire County, to create the Derwent Valley Water Board in 1899. This brought about 'joined-up thinking' in the development of the region's water resources and remains a model of local authority collaboration today.

Plans to construct reservoirs in the Derwent Valley came to fruition in 1912 when Howden Reservoir was completed, although Nottingham did not use the water until 1917, due to quality issues. Ladybower followed in 1945, and Derwent in 1960. Five borehole stations were added between 1945 and 1969, and steam engines were replaced by electric pumps in the 1960s. A new works and reservoir at Church Wilne, which includes treatment with activated carbon, was completed in 1967 and has been twice extended since. The long-planned new reservoir at Carsington took until 1992 to complete.

In 1974, water supply and sewerage ceased to be the responsibility of the City of Nottingham, and it became part of the remit of the Severn Trent Water Authority. Following privatisation of the water industry in 1989, the responsibility passed to Severn Trent Water, one of ten water and sewage companies in England and Wales.

WWI as a Watershed

The first World War was a turning point in both politics and public health as David Lloyd George declared that the country would build "Homes fit for heroes" in 1918. This was not just electioneering as, when young men turned up to volunteer for the army, many failed the basic medical as their poor living conditions led to them being unfit. What followed was a council house building programme which continued into the 1970s. The private sector also responded as virtually all terraced house built after 1920, had a bathroom whilst the earlier ones didn't. This is why you see so many extensions at the back of early terraced houses – it's a bathroom and sometimes a kitchen as well – no more tin baths in front of the fire!

Case study – The Black Country Sewage Farms

In the intervening period, between the development of water supply systems and the corresponding sewage treatment plants, many areas resorted to spreading sewage onto land for natural treatment. This was not quite as altruistic as it seems as they used it to irrigate crops. The land owner – some say it was Lord Derby - developed a sophisticated system to transport the sewage to the farms under gravity. Much of the western side of the Black Country is at an elevation which allows water to gravitate some way westwards into the countryside. He had a collection pond (The Pound) built as a header tank for a pipeline which took the sewage out to the farms some miles away. The system was still in use up to the 1970s when a treatment plant was built at Lower Gornal. The Severn Trent plant at Roundhill is located in one of the historic farms.

Sewage farms remained the main form of treatment until Parliament passed legislation requiring more sophisticated methods but they are still common in warmer climes where sewage is seen as a valuable resource for both irrigating and fertilizing crops.

Rural Systems

After WW2 there was a massive push to improve sanitation in the rural areas which generally relied on their pipes being connected directly into the village drain. Most Rural District Councils did not employ engineering staff and so employed consultants to design collection and treatment systems. After some debate about the relative merits of localised or centralised treatment, most opted to build a single plant and pump the sewage there from surrounding villages. If you travel through the countryside, look at the topography and seek out the lowest point in a village. Usually, you find a small building there with an electricity supply. And if you have a sensitive nose, you may sniff out its purpose.

Many of the earlier pump stations were constructed to a standard design with dual wells and a vertical shaft drive between the motor and the pump body. Most of these have since been replaced by modern submersible pump stations which are reliable, efficient and easier to maintain.

The Slum Clearances and New Towns

During the 1960s there were two parallel initiatives which had a major impact on the way we lived. All of the major cities had slum clearance programmes which were aimed at replacing 'Coronation Street' style terraces with modern housing. Unfortunately, many were destined as walk-up flocks of blats (excuse the Spoonerism) and because of their inherent design faults, many have since been demolished. Whilst the supply of clean water was covered by the 1945 Water Act, the situation with drainage systems was more complex. Many cities would have liked to convert to separate systems, but there was often no obvious outlet for the storm sewers and so the combined system was perpetuated.

Whilst the slums were being cleared, there was a need for people to move out of the cities and so 'overspill' estates were built. These were nearly always on green-field sites and so separate systems of drainage were constructed. Where major relocation of population out of the cities was essential, 'New Towns' were planned and predominantly on green-field sites. They all featured separate systems of drainage.

The 1974 Reorganisation

The 1973/74 local government reorganisation In England and Wales included the setting up of the ten Regional Water Authorities (RWAs) which were somewhat like local government but supposedly independent of it. It's acceptance by major cities like Birmingham and Manchester was only brought about by the appointment of local authority representatives to the new boards which ran them. In addition, the private water-only companies remained independent. Also, sewerage functions (along with minor highways) were delegated to local councils under so called 'agency agreements' which enabled them to keep engineers employed in council offices.

Wales was covered by a single authority except for the small area within the Severn catchment which was included in Severn Trent's remit. The Elan Valley reservoirs, which are at the head of the Wye catchment, were, thus, transferred to Welsh Water, much to the chagrin of the City of Birmingham who took the Minister to the High Court over it.

The functions of the River Authorities were subsumed into the new authorities and they were given the tasks of regulating water quality in rivers and controlling abstractions. The Drinking Water Inspectorate continued its role of monitoring potable water quality.

Scotland and Northern Ireland were included as single entities under the control of a Ministry.

The 1989 Privatisation

England and Wales underwent a further change when the Thatcher government decided that privatisation was the best future for the industry which had been starved of capital funding as an arm of government. Shares were sold to the public in line with the other utility share offers and the Company Boards were to set about improving their services in line with new legislation. The economic regulatory functions were transferred to the newly created Water Services Regulatory Authority (Ofwat) and environmental regulation to the National Rivers Authority which then morphed into the Environment Agency.

Scotland and Northern Ireland were not included in the privatisation.

2012 The Canal and River Trust set up

In 2012, the government decided that it did not want the financial burden of managing rivers and canals and so set up The Canal and River Trust to manage all navigable waterways. However, they left the responsibility for water quality with the Environment Agency as well as responsibility for non-navigable main rivers.

The Future and Climate Change

Climate change has brought new challenges which will need to be addressed as recurring floods come into our living rooms care of TV coverage. Whilst these provide the media with great material, it is the imminent possibility of water shortage in London and the South East which poses the biggest problem. Whilst there is much talk there is little sign of positive action to address this threat.

Micro-plastics

Our modern toys are designed to appeal but much of them are made from plastics. What happens to all of our waste plastic, as it ages, remains a big question. Whilst we are largely aware of most other threats to the aquatic environment, the issue of micro-plastics remains to be solved.

Colourful plastic toys by Shankar S

The Current UK Water Industry and Regulation

What does the UK water industry look like now? The underlying structure is almost unchanged from that brought about in 1974 though, following privatisation, ownership of the companies has gone through many changes. Nowadays water companies are treated almost like commodities either on the Stock Exchange or by large investment firms. How this is viewed depends largely on your politics and your view of capitalism.

Whilst Scotland and Northern Ireland continue to keep their water business in public ownership, with Wales in its own set-up, England's big nine remain roughly the same. The table which follows attempts to put things into context though the figures are not always easy to obtain. Why Ofwat does not have a standardised table beggars belief and United Utilities' failure to tell us how many people they serve is difficult to understand.

Other bodies such as The Canal and River Trust are not shown as they are not party to this system of regulation. Details of the regulation system are contained in Appendix 6.

England and Wales

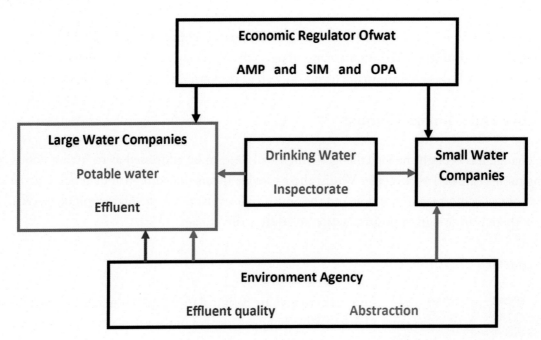

AMP = Asset Management Plans (a new one every five years)
OPA = Overall Performance Assessment (comparison with other companies) which was replaced by:
SIM = Service Initiative Mechanism which was replaced by C-MeX and D-MeX in 2020

Which all goes to gives the appearance of a job-creation-scheme at Ofwat.

There is also an informal body: The Consumer Council for Water (CCWater)

Scotland

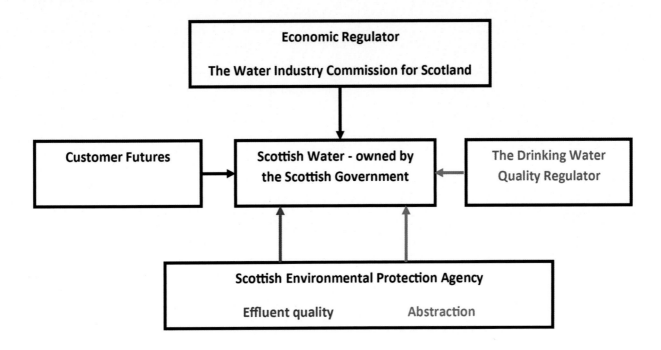

Northern Ireland

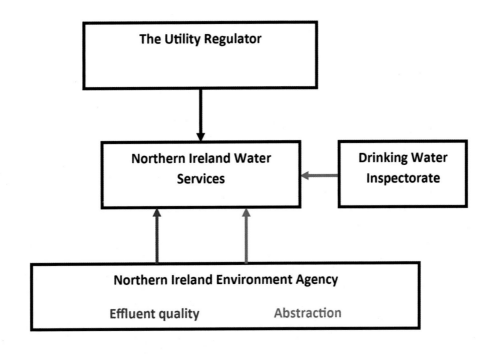

Company	Owners	Headquarters	Turnover £m/a	Employees	Water customers (k)	Sewage customers (k)
Anglian Water	Consortium	Huntingdon	1,270	5,000	6,500	6,400
Dwr Cymru	Glas Cymru NfP	Cardiff	780	3,500	3,000	3,000
Northern Ireland	N I Government	Belfast	540	1,310	1,590 est	1,320 est
Northumbrian*	Chung Kong et al	Durham	760	3,110	2,700	2,800
Scottish Water	Scottish Government	Dunfermline	1,670	4,280	5,450	5.450
Severn Trent	Shareholders	Coventry	1,830	6,170	8,000	8,000
South West Water	Pennon Group	Exeter	550 appx.	1,500	1,700	1,700
Southern Water	Greensands	Worthing	780	2,090	2,500	4,600
Thames Water	Kemble Water	Reading	2,100	6,300	9,000	15,000
United Utilities	Shareholders	Warrington	1,810	5,000 est	6,000 est	6,000 est
Wessex Water	YTL Corporation	Bath	510	2,380	2,800	2,800
Yorkshire Water	Kelda Group	Bradford	1,100	3,500	5,000	5,000
Affinity Water	Alianz group	Hatfield	400	1,300	3,600	0
Bristol Water	ICON	Bristol	120	400	1,200	0
Hafren Dyfrdwy	Severn Trent	Wrexham	20	130	87	20
Portsmouth Water	Shareholders	Havant	40	260	700	0
Bournemouth Water	Pennon	Bournemouth	60	70	500	0
South East Water	Hastings	Snodland	248	950	2,200	0
South Staffs Water	Aquainvest	Walsall	125	440	1,700	0
Sutton and E Surrey	Sumitomo	Redhill	66	350	735	0

*Includes Essex and Suffolk Water

As Ofwat does not provide a definitive table to fulfil this requirement, the figures quoted above which relate to 2016, are for guidance as to the size of the company only and should not be relied upon for any other purpose. There are a few very small companies which, whilst technically independent, are managed by the larger companies. Jersey Water, Guernsey Water and Isle of Man Water are all government owned but are not part of the United Kingdom.

Water Legislation in the UK

Not long after I had started work, my boss explained to me that, as an individual, I could do almost anything that I wanted to do unless there was a law which forbade it. In contrast, a local authority can only do things that they are empowered to do explicitly, by law. This principle carried over to the utilities which now run the water industry.

Water is free for all – isn't it? In many countries their legislation contains a right to free water and many argue that it's a fundamental human right. The law in Guyana, allows anyone to access a source of water and take away fifteen litres per day for their own use, however, this does not include it being treated and delivered, under pressure, direct to your house or business - for that you have to pay your water bill which depends on the quantity recorded by your water meter. The debate continues but water companies argue that it's not their job to deliver free water. If it's to be supplied free-of-charge, then that's a job for politicians to decide and to then to provide a mechanism which covers the cost.

What follows is a list of most of the more important pieces of legislation to affect the wider industry. Scotland and Northern Ireland have their own laws which generally match those of England and Wales. Most of the earlier acts have been absorbed or amended by the later ones.

1848 Public Health Act

Set up Local Boards of Health

1866 Sanitary Act (aka The 1866 PHA)

Allowed the formation of drainage districts and enabled the provision of better house drainage

1875 Public Health Act

Local authorities to purchase, repair or create sewers; control water supplies; regulate cellars and lodging houses and establish by-laws for controlling new streets and buildings. Original definition of 'sewer'

1870 Education Act

Education to include heath issues

1872 Metropolitan Water Act

Unified the management of London's water supply which had previously had eight separate companies, each with its own standards

1919 Housing Act

Homes fit for heroes and council houses

1930 Reservoirs (Safety Provisions) Act

Design, construction and regular inspection of reservoirs

1930 Land Drainage Act

Setting up of the Internal Drainage Boards

1936 Public Health Act

Set up The General and Local Boards of Health. Provisions with regard to definition and adoption of sewers and sewage treatment, etc.

1937 Public Health (Drainage of Trade Premises) Act

Control of trade discharges to sewers

1945 Water Act

The 'Waterworks Code'. Power to make bye-laws

1947 Fire Services Act

The supply of water for fire fighting

1948 Water Act

Amendments to the 1945 Act

1950 Public Utilities Street Works Act (PUSWA)

Control of pipe and cable laying in streets. The 'Street Works Code'.

1951 Rivers (Prevention of Pollution) Act

Prevention of pollution

1960 Clean Rivers (Estuaries and Tidal Waters) Act

Pollution of coastal waters and estuaries

1961 Public Health Act

Amendments to 1936 Act; trade effluent consents

1961 Rivers (Prevention of Pollution) Act

Prevention of pollution

1963 Water Resources Act

Formation of River Authorities in 1965. Control of abstraction

1968 Water Resources Act

Amendments to the 1963 Act, repealed in 1991

1971 Water Resources Act

Regulation of river flows

1973 Water Act

Reorganised the water industry by setting up Regional Water Authorities

1974 Control of Pollution Act (COPA)

Controlled waste; discharges of polluting or solid matter; discharges from mines; trade effluents and sewage effluents

1974 Health and Safety at Work Act (HASAWA)

Health and safety duty of employers

1975 Reservoirs Act

Replaced the 1930 Act and required counties to keep a register

1975 Salmon and Freshwater Fisheries Act

Maintenance of freshwater fisheries; fishing licences and water bailiffs

1976 Drought Act

Reaction to the 1975/76 drought.

1976 Land Drainage Act

Consolidation of land drainage powers and responsibilities. Land drainage committees and 'main river'

1977 Water Charges Equalisation Act

Power of the Secretary of State to order a levy to level up charges; repealed 1983

1977 Town and Country General Development order

Permitted development

1989 Water Act

Privatisation of water companies

1990 Environmental Protection Act (EPA)

Controlled waste; contaminated land; litter; 1991 Waste Management Licensing Regulations

1991 Water Industry Act

Consolidation of earlier acts and consultation on fluoridation schemes

1991 Water Resources Act

Water resources, water pollution and flood defences

1994 Waste Management Regulations

The carriage and disposal of all types of waste

1997 Confined Spaces Regulations

Regulated working in confined spaces

1999 Water Industry Act

Forbade disconnection for non payment of water bill

2003 Water Act

Abstraction, impounding and navigation

2012 Transfer of Assets

Set up the Canal and River Trust

European Community Legislation which affected the UK

91/271/EEC - The Urban Waste Water Treatment Directive (banned the disposal of sewage sludge at sea)

98/83/EC – The Drinking Water Directive

2000/60/EC – Water policy

2006/118/EC – The Groundwater Directive

2007/60/EC – The Floods Directive

2008/105/EC – Environmental quality standards in surface waters

Generally, the EU directives are converted into legislation by the member country. Whilst the basic requirements are quite clear in the original, subject to translation in some countries, once the civil service get hold of it anything can happen as they have to tailor it and then amend all of the existing laws and point out how it applies. The 1994 Waste Management Licensing Regulations are a case in point as 'wherewithals' and 'notwithstandings' proliferate.

There is a hierarchy in law which could be said to approximate to this:

- Fundamental common law and human rights
- Primary legislation - laws passed by Parliament
- Regulations based on enabling law
- Interpretations of the law in the courts ('precedence')

For those of you who find British law to be somewhat complicated and confusing, this is what Charles Dickens wrote in his masterpiece *Bleak House*:

"The one great principle of the English law is, to make business for itself."

Law by Woody H1

Water Chemistry and Properties

The first thing about water is that it's very reliable and found almost everywhere on the planet. That's useful if we have systems of measurement and want to relate them to each other so how does volume relate to weight (not worrying about the difference between mass and weight)? Well, to put it simply – a litre of water weighs in at (almost exactly) one kilogram and almost everything else, except time and temperature, follows from that. When we get to temperature then we have to consider how to set that up and Swedish astronomer, Anders Celsius proposed that we divide the scale between its freezing point and its boiling point into 100 units which were called 'grades' at one time – hence the original name for it as the centigrade scale. So freezing water is at 0°C and boiling water at 100°C and we now call it after Celsius. See the later chapter on units.

Photo by Fox-kiyo

As most of this book is concerned with water in its liquid state, we won't spend too much time on its solid (ice) state and its gaseous (steam) state although it might be worth mentioning that water increases in volume when it freezes, hence the bursting of pipes during heavy frosts.

Solutions

Water is versatile in that it can adopt many different properties by absorbing other chemicals. When it does this we call the result a 'solution' as the other chemical is 'dissolved' in the water. It may, therefore, become a drink, a food, a cleaner, a healer, a conveyor and many other things.

Oxidation/Corrosion/Rust

Water's tendency to dissolve things into a solution is a problem as it is one of the commonest sources of corrosion. Where water is present, with dissolved oxygen, it will corrode many materials but particularly metals including iron and steel. Whilst the addition of water to cement is part of the process for making concrete and mortar, they do not appreciate constant exposure to it.

Hydraulics and Compressibility

Water is comparatively incompressible and so it has given rise to a whole area of technology which uses it to transfer power. Up until the 1960s, Manchester had a 'Hydraulic Power Station' with a system of high-pressure pipelines under the streets. The high-pressure water was used to power machinery and a number of hotels in the city centre used it to raise their lifts. Whilst the basic word appears to be based on water, the term 'hydraulics' has come to mean that which is connected with high pressure fluids – hence 'fluid mechanics'.

White mountain waters streaming among rocks by Horia Varlan

Laminar and turbulent flow

In fluid mechanics, flowing water behaves quite differently according to whether the flow is laminar or turbulent. Whilst the difference is obvious to most casual observers, technology requires a specific measurement and this is determined by the 'Reynold's number'.

Cooking

Roasting, grilling and boiling are perhaps the commonest forms of traditional cooking and it is boiling that requires water to make it happen. Whilst the tenderisation of the ingredients is the object, the killing of bacteria is also crucial. Microwave ovens have become very popular due to their speed and efficiency; they use an electromagnetic source to jiggle the water atoms which increases their temperature and hence cooks.

Cooking is hard work by Mripp

Cleaning

Cleaning agents may be either natural or synthetically developed and are generally classified as: water; detergents; abrasives' degreasers; acid cleaners; organic solvents; etc. Water is the commonest cleaning material in the world, especially when it contains soap or detergent in solution. The soap molecules encourage dirt particles to adhere which make them easy to separate from the article which is being cleaned. Some say that sand, as an abrasive, is the next commonest cleaning agent after water.

Drinks

Most drinks are water based and have a flavouring added in solution. Carbon dioxide is added to make fizzy drinks and sugars to make energy drinks. What we call 'alcohol' is mostly water with some concentration of the intoxicating stuff. Most beers are now around 5% alcohol whilst wines vary between 8 and 13%. Fortified wines are typically around 20% whilst spirits come in at, generally, 40%. This latter concentration is due to the property which means that distilled alcohol evaporates off at a 'constant boiling mixture' which is around 60% water and 40% alcohol. We have a short chapter on the drinks industry later in the book.

Coca-Cola by Twm1340

Diseases

Perhaps the best known water born disease is cholera but there are many others and simple food poisoning is probably the most common. They can be caused by protozoa, bacteria, viruses, algae or parasitic worms. Cryptosporidium, amoeba, giardiasis, botulism, e coli, dysentery, legionella, salmonella, polio, typhoid and typhoid fever can all be spread through contaminated water along with some forms of hepatitis. Whilst most are sourced from infected individuals cryptosporidium can be present in the treatment process itself. Acute gastroenteritis is more a statement of the symptoms rather than the cause. WASH (water, sanitation and hygiene) is the way to avoid problems which can be fatal.

Travellers' Tummy

Along with food poisoning this is probably one of the commonest forms of gastro-intestinal upset. It may be caused by an infection but also just by drinking water that you are not used to. 'Berkshire Belly' was common in this country up to the 1970s but less so nowadays. When travelling abroad, it's good practice to have some Imodium with you just in case. The best known forms include: Montezuma's Revenge; Madras Ass; Delhi Belly, Hong Kong Dog, Casablanca Crud, Tokyo Two-step and the Kathmandu Quickstep. We really should have another one - 'Death on the Nile' for when you visit The Pyramids.

It can be wise to have your own precautions available if you are not sure about the water. Drinking beer, wine or spirits should be OK as the big drinks manufacturers take stringent measures to ensure that their product are safe. You can use Milton or sterilising tablets to sterilise water that you are not sure about and, in emergency, even a very small amount of domestic bleach. Peroxide has also been used. If you can smell the chlorine in the swimming pool, it should be safe but otherwise, stay away.

Swimming pool in the Sky by Green Kermit

Legionella

Legionella came into the news in 1976 when an outbreak of Legionnaires' disease killed 36 people at a conference in Philadelphia. It is mainly spread from poorly maintained cooling systems but can be present in any water system where there is a water/air surface. Ultra violet radiation is the best way to maintain the water and avoid the disease.

Peroxide

Water has a near twin which is hydrogen peroxide. Exactly as its name suggests it consists of equal numbers of hydrogen and oxygen atoms which make up H_2O_2. Like chlorine, peroxide is a powerful oxidising agent and thus can be used as a cleaner and as a disinfectant.

Chlorine

Strangely, chlorine is also classed as an 'oxidising agent' which makes it the preferred chemical for keeping water pure. When in solution it rapidly combines with pollutants and micro-organisms, killing them almost instantly. When in solution, it evaporates slowly into the atmosphere if given the chance and it is this smell which we tend to associate with purified water. Following treatment, it is normal practice to dose with chlorine gas or a chlorine salt to achieve around 1.0ppm in solution. The aim is for this to kill off any remaining organisms which could cause illness and a residual concentration of 0.1ppm is normally aimed for.

Electrolysis

Water can be used as a source of both of its constituent gases. If battery electrodes are placed in water and energised, then hydrogen is given off at the cathode and oxygen at the anode.

Hydrogen and Oxygen Electrolysis by epredator

23

Hardness

'Hard' water has high mineral content in contrast with 'soft' water. When water percolates through deposits of limestone, chalk or gypsum, which are largely made up of calcium and magnesium carbonates, bicarbonates and sulphates, some of them dissolve in the water making it hard. This makes it difficult to use for cleaning without lots of soap or detergent. Many use water softeners in the home to correct it. Consuming hard water is said to be better for your health.

Stalagmites In Poole's Cavern by AndrewJW

Nitrates

The most common cause of methemoglobinemia or 'blue baby syndrome' is through the ingestion of nitrates through well water or foods. Most nitrates in potable water come from agricultural fertilisers which have not been taken out in the treatment process.

Water Supply

Meaning of Potable

Taken from the Latin verb 'potare', which means to drink, it's exactly what it says on the tin (did we say that somewhere else?) In practice it means water that has been abstracted and treated to drinking water standards though these may vary from country to country. The most reliable standard is that compiled by the WHO though most countries will aim at higher standards.

Domestic Consumption

There are many sets of statistics but, strangely, no government department or regulator appears to produce a standard set which can be used for guidance and analysis of trends. This would be useful, when one is being bombarded with advice on how to reduce your water consumption and hence carbon footprint to have, at least some idea, of where the water goes. Here is just one précis of stats in litres per capita per day (l/c/d) which can be found on the web:

- Toilet (loo) 45 [32%];
- Bathroom (bath and shower) 24 [17%];
- Utility (washing machine) 17 [12%];
- Kitchen/cooking 50 [36%];
- Garage/car/patio 4 [3%]

making a total of 140 [100%] compared with the normal total quoted of 142 l/h/d. Would be nice if they used a consistent format – wouldn't it?

'Water Supply' in its Broadest Sense

The term 'water supply' has two related meanings. In its broadest context it covers the whole gamut from abstraction to the customer's tap. But, for those involved in the industry, it can also mean that part of the system which comes between 'treatment' and 'distribution.'

The supply of water to customers takes place in three distinct, but connected 'sections'. We start with raw water which is abstracted, from rivers, aquifers or reservoirs, and then treated. It is then 'supplied' in large diameter mains called 'trunk mains'. 'Transmission' is the term applied to the transfer of bulk water from a water source to a treatment plant and from a treatment plant to a large consumer or a town or city. This transfer of large quantities of water takes place in 'trunk' mains so-called because years ago hollowed out tree trunks were used as water pipes. This also explains why the word 'branch' is used for connections to the trunk main.

Branches take the water from the trunk main and distribute the water to 'customers' (sometimes called 'consumers' – if you want to know the difference see the Finance chapter – Billing and Collection). These branch pipelines are called 'distribution mains'. Trunk mains are large diameter mains and in the UK can be up to about 2.5 m diameter. However, in a rural area even a 150 mm diameter main could be considered as a trunk main as it may take water from a local water source to a village. In general it is reasonable to consider mains as 300 mm (12 inches) and above as trunk mains and smaller mains, from 80 mm (3 inches) to 250 mm (10 inches) diameter as distribution mains.

The third category of supply is the thousands of small 12mm (half inch) pipes, though modern connections tend to be 25 mm (one inch), polyethylene pipes which connect the customers' properties to the water companies distribution pipes. For legal and practical purposes service pipes are separated into three distinct categories: 'communication', 'supply' and 'service' pipes.

The service pipe is connected to the distribution main by a ferrule which, depending on the pipe material, can be threaded directly into the pipe or via a saddle to provide a strong, watertight connection. The length of pipe between the water main and property boundary is called the communication pipe and it is the responsibility of the water utility company to maintain and repair it including the water meter if fitted. A stop tap (sometimes called a 'stopcock') is fitted at the boundary of the property to enable the customer to turn off the water. The continuation of the service pipe after the stop tap is called the supply pipe which feeds water to the cold water system within the property and is the responsibility of the property owner.

A Diversion

I was travelling to Skye with a friend and we and decided to break our journey in The Trossachs. Arriving early afternoon, we found that we had time to take in Loch Katrine aboard the steam yacht Sir Walter Scott. As the guide told us about Glasgow's water supply, I enjoyed one of the Katrine beers which they serve aboard. He told us that the beer was brewed in Glasgow from Loch Katrine water and hence it had come full circle.

Steam Boat Sir Walter Scott" by Marsupium photography

When I set about writing this book, I decided to check up on the facts and found that the Tryst Brewery, which made the beer was actually in Falkirk. I contacted them to ask what water they used but got no reply. My assumption, therefore, is that the water used to brew the Lock Katrine beers is actually from Falkirk's own source which is the Carron Valley.

On the way back we went via Falkirk to see The Kelpies and the Falkirk Wheel which links the Forth and Clyde Canal with Scotland's Union Canal.

Falkirk Wheel by M McBey

Water Resources

The UK has plentiful rainfall to serve its needs but, unfortunately, it doesn't all fall in the right place. Wales, Scotland and Northern Ireland are all self-sufficient as is the North of England and the South West. The Midlands, East and South East are in deficit which creates a potential serious problem in the long term. We will avoid the temptation to describe the issues in numerical terms and will concentrate on descriptive means. When we talk about 'resources' we are generally referring to water before it is treated to make it 'potable' – so it's often referred to as 'raw water'.

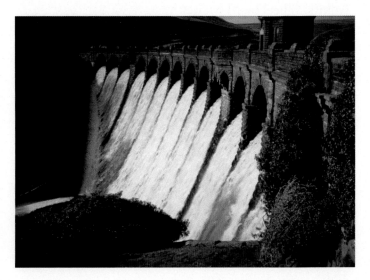

Craig Goch in the Elan Valley

Prior to Victorian times, most water was abstracted locally, from streams, rivers, lakes and aquifers but such resources are soon overloaded by the needs of a growing population. When you rely on local sources and that gets over-abstracted, or polluted, then you have a major problem, especially in the days before we had bottled water to drink. The cholera epidemics of the mid-nineteenth century brought the problems into sharp focus.

During the Industrial Revolution, the need for clean water ballooned due to the growth of the cities and water-hungry manufacturing. Thus Victorian engineers were appointed to look for means to bring about a transformation in the supply of clean water and much of this infrastructure remains with us today – indeed we would be in dire straits without it.

Generally, we take around a third of our water from lakes and reservoirs, a third from rivers and the remaining third from aquifers. Let's look at Birmingham which as "The City of a Thousand Trades" grew exponentially from a small village to become England's second city (Manchester and Liverpool both dispute this title) in a relatively short time. City father, Joseph Chamberlain (parent of future Prime Minister, Neville) is credited as being the man who was responsible for the City's transformation and the development of its water supply. Engineer, James Mansergh, was engaged to find a solution to the City's problem and he found one in Central Wales at the head of the River Wye in the Elan catchment. Whole villages were removed when the dams were built, under an Act of Parliament, along with an aqueduct taking the water, by gravity, all the way to Frankly Reservoir where a treatment works was built.

In the North West, Manchester ("Cottonopolis") and Liverpool had similar problems, though the latter had very little in the way of industry. The cotton industry in the North West, and Manchester in particular, was very water intensive and so new sources were essential to the growth of that industry along with manufacturing at such major centres as Trafford Park. Manchester Water Department looked north and went to The Lake District for it new sources. Haweswater and Thirlmere were used as sources and two major aqueducts were built to bring the water south. Liverpool looked in the opposite direction and built a new aqueduct to take water north from a new reservoir at Vyrnwy supported by Lake Bala.

Kent Mill, Chadderton by Chris Allen

All of the major cities built new infrastructure to support their water shortages in late Victorian times and the city based water departments expanded their remit to serve most of the adjoining areas. They were later to become the skeleton of the Water Authorities which were created in the 1970s.

So, looking at what the Victorians did, we find much common ground. If we have a major centre of population, which has an excessive need for water, then we need to look elsewhere for a new source. This may be an existing river or lake but, if there is not a suitable intake, then we need to build dams to support a reliable supply. Getting the water from the source to the city then needs an aqueduct to transfer it which may be pumped or, preferably, by gravity.

In more recent times, the Cities of the East Midlands got together to build the Derwent Valley system which has been modified by the addition of Ladybower Reservoir which also serves Sheffield. Massive new reservoirs were added at Scammonden, Kielder and Rutland, all of which are open to the public and well worth a visit. The last major dam to be built in England was Carsington which is linked into several of the Midlands' supply systems.

Valve tower on Kielder reservoir by Peter Moore

Groundwater

Nottinghamshire was, once upon a time, famous for its coalfield and a large proportion of the county has been mined due to the quality of the coal and its accessibility. This aspect of its geology has also resulted in its bedrock – the Nottinghamshire sandstone, having an aquifer of great potential for water supply which has been exploited in contrast to adjoining municipalities who built reservoirs. The big advantage of using groundwater is that it requires little or no treatment other than chlorination.

The steam-powered pumping station at Papplewick is now run by a charitable trust and has open days throughout the year when the beam engine can be seen running.

Obviously, we rely on more modern methods nowadays and the current borehole can be seen in the grounds with a very much smaller footprint.

Papplewick Pumping Station by Amanda B H Slater

Many small water companies, especially in the south of England, rely on groundwater for their whole resource and this is a constant subject of debate due to their effect on depleting the chalk streams upstream of their localities.

Confined and Unconfined Aquifers

Confined aquifers have an overlying protective layer, usually of clay or other impervious soil. This confining layer offers protection from surface contamination especially from pesticides and agricultural fertilizers.

Unconfined aquifers have no protective layer at the surface and can be contiguous with the groundwater, making them susceptible to contamination. This means that the surface area has to be protected - subject to the control of chemicals used in farming and other industries which might pollute the water.

A 'perched' aquifer is one which has a lens of water which lies between layers of dry geological formations.

Case Study – The River Severn as a Water Resource

The River Severn (or Sabrina/Hafren as it is sometimes known when endowed with character) is not just a land drainage channel as it also serves as a navigable waterway and a source of raw water for the communities along its route. It's use as a water resource has developed considerably over the years and it is now managed to a very great extent. The weirs, which were built to aid navigation, also serve as devices to measure the flow. Strangely, Birmingham gets most of its water from the adjoining River Wye catchment due to the City fathers' decision to develop the Elan Valley dams. Water quality (at that time) and the lack of pumping would have contributed to their decision making. Trimpley was built much later as a back-up to the Elan Valley supply. The Severn is now navigable as far up as Stourport-on-Severn but, in the past, boats went as far north as Bewdley.

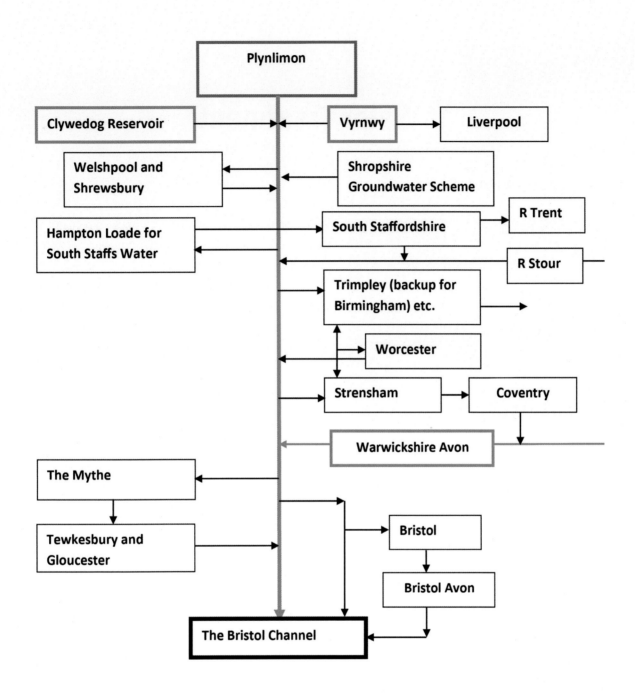

Water Treatment

The WHO and British Standards

Water treatment is intended to make water potable – i.e. drinkable. This requires fairly complex systems which gradually refine the water until it is considered 'pure' and safe. The World Health Organisation sets basic standards for 'safe' water and these often form the basis of an individual country's standards. The UK had its own standards before the WHO and these have often formed the basis for other countries, especially those in the Commonwealth.

Melbourne Water Treatment Plant in Leicestershire.

The UK Standard is now based in that of The European Union and the following parametric standards are included in the Drinking Water Directive which are expected to be enforced by appropriate legislation in every country in the Union. The following truncated list, which is based on that in Wikipedia, indicates the sort of levels that are allowed to be present:

- Cadmium 5.0 µg/l

- Chromium 50 µg/l

- Copper 2.0 mg/l

- Cyanide 50 µg/l

- Fluoride 1.5 mg/l

- Lead 10 µg/l

- Mercury 1.0 µg/l

- Nickel 20 µg/l

- Nitrate 50 mg/l

- Nitrite 0.50 mg/l

- Pesticides 0.10 µg/l

- Pesticides - Total 0.50 µg/l

Bank-side Storage

Many plants which take their water from a river have 'bank-side storage'. This is a relatively small reservoir which is situated between the river intake and the treatment plant. It provides a degree of storage and some settlement which aids the treatment process. It also protects against pollution events in the river.

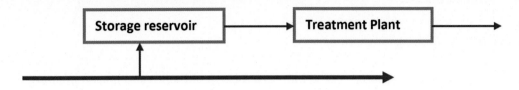

Intake and Screening

The first part of the treatment process is the 'intake' which is a relatively simple mechanical process. Water is abstracted from its source, either by gravity (such as a draw-off tower from a reservoir) or by pumping. If it's from a river, it is then usually screened – firstly for coarse material and secondly through fine screens to remove floating debris.

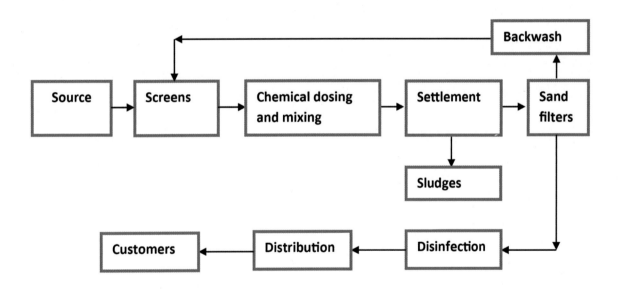

Pre-chlorination

Some processes involve adding chlorine to the water before its treatment in order to kill off most of the harmful organisms. If the water is relatively clean then this is not undertaken. Pre-chlorination would require that the water be taken through a tank or, better, an elongated channel in order for the chemical to have time to act.

Settlement/chemicals

A chemical such as aluminium sulphate (alum) or iron sulphate (ferric) is added to aid settlement and mixed to ensure an adequate contact time. The mixture is passed through a settlement tank which is designed to make the resulting sludge settle out as a floc (flocculation). Some processes use a 'flash mixer' with a polymer to aid settlement.

Sludge removal

The resulting sludge, which is quite thin, is removed at the bottom of the tank and taken away for thickening or disposal. As waterworks sludge contains the chemical used to settle it, it is suitable for disposal direct to a foul sewer which will take it the sewage treatment plant. Here it actually aids the settlement of the sewage sludge which is much thicker.

Sand filtration

The settled water is then passed through a sand filter to remove small solids and most remaining organisms. The two main types of sand filter are 'slow sand' and 'rapid gravity' filters. A slow sand filter works with a biofilm on the grains of sand which trap particles. It has a low energy profile and only needs occasional replacement of the sand. A rapid gravity filter uses the sand directly as a filter medium and requires 'backwashing' at

regular intervals using compressed air to remove the deposit that has settled. The backwash water may be treated or returned to the intake.

Bedfordshire sand is the preferred medium for sand filters and Leighton Buzzard sand has even been exported to Saudi Arabia and Egypt.

Activated carbon may be used in conjunction with the sand filters to remove chemicals and taste.

Dissolved Air Flotation (DAF)

An alternative to the use of a settlement tank is to reverse the process by making the solids lighter than the water which makes them float. After addition of the chemical, the water passes through another tank which has fine bubbles of air fed in at the bottom. These bubbles attach themselves to any floating material and carry it to the top where it is overflowed via a weir. The clarified water is then passed to a storage tank where chlorine is added to the desired concentration. A contact tank or channels may be used to ensure that the process is complete.

Ultra violet

Some plants now use modern methods of final clarification such as ultra violet. The water is passed through a channel and exposed to UV light which kills any remaining pathogens.

Disinfection/chlorination

When the water has been finally treated, chlorine is dosed to ensure that any bacteria in the distribution system are killed off. An alternative is to use ozone but a residual dose of chlorine will still be required.

Hardness, softness, heart disease

Hardness in water is caused by dissolved calcium carbonate and occurs, especially in groundwater in chalk areas. It makes the water 'hard' which means that you have add a lot of soap to make it soft enough to use in washing. Most people prefer 'soft' water to drink despite 'hard' water being healthier. Statistics show a lower incidence of heart disease in hard-water areas.

Fluoride/teeth

Currently, around six million people in England live in areas with water fluoridation schemes, mainly in the West Midlands and the North East. Birmingham was the first permanent scheme to commence in 1964. It is estimated that, around four hundred million people in some twenty five countries are currently served by water fluoridation schemes. There are an additional fifty million people, worldwide, consuming water with naturally occurring fluoride at or around the same level as used in fluoridation schemes.

Photo by The Lamb Family

It's been said, but I can't confirm it, that the River Cole, which drains parts of Birmingham, has a background concentration of fluoride which is measurable. The only likely source would be leakage of potable water from the city's supply system.

pH Correction and Lime Dosing

If the final water has an acid nature then an alkali (usually lime) may be added for pH correction. In areas with lead pipes, then lime may be added to reduce plumbo-solvency in the pipes.

Nitrates

If nitrates are still present in the treated water then ion exchange or membrane treatment may be required. This removes the risk of 'blue-baby syndrome'.

Endocrine Disruptor Chemicals

Recent studies have shown that EDCs, which persist in wastewater and through into potable water, can cause adverse effects in animals and even humans. EDCs in the environment have been related to reproductive and infertility problems in wildlife and restrictions on their use have been associated with the recovery of some wildlife populations.

Cyanobacteria (blue-green algae)

Aquatic cyanobacteria are known for their extensive and highly visible blooms that can form in both freshwater and marine environments. The blooms can have the appearance of blue-green paint or scum and can be toxic. They occur where there are high concentrations of nitrates and phosphates in the water.

Groundwater Treatment

Most ground waters have passed through the soil and then through sub-soil and often through rock such as sandstone. Sandstones and chalks provide excellent filtration and do not naturally harbour harmful bacteria. This means that many ground waters can be put into supply, without further treatment after abstraction, only requiring chlorination.

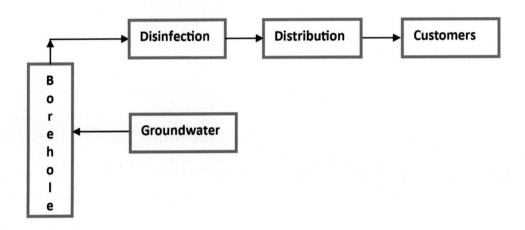

Reverse Osmosis – Case Study – Beckton Desalination Plant

This is a relatively modern process whereby water is forced, under a very high pressure through a fine membrane which prevents even very small particles from passing through. The process is commonly used for 'desalination' which removes even the dissolved salt in seawater. It is roughly three times the cost of normal treatment due, in part, to cost of pre-treatment which is close to potable standards.

Beckton is also the site of London's major sewage treatment plant on the north bank of the Thames estuary. When London was chosen as the venue for the 2012 Olympics, Thames Water predicted a possible shortage of potable water in the capital and so set about finding a solution. They decided to build a desalination plant near Beckton which would treat brackish water from the Thames estuary to a potable standard. It's said that it would actually have been somewhat cheaper to take effluent direct from the sewage works as that would not have contained salt. The decision to treat the estuarine water which contains Beckton effluent anyway, was based more on 'the yuk factor' rather than economics.

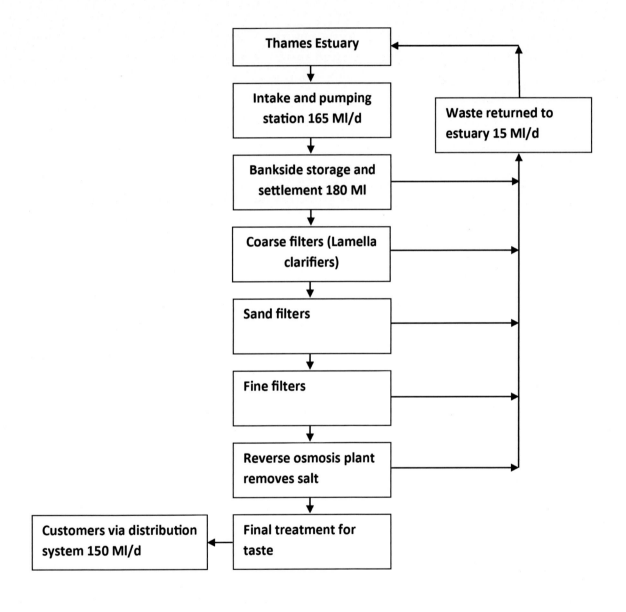

Water Supply and Storage

Whilst some use the term 'supply' to apply to the entirety of the systems (as in the figure below), I use the term 'supply' to refer to the movement of treated water from the treatment plant to the area where it is needed. Generally, it does not involve any connections to an area through which it passes other than to the 'distribution' systems. We also use the term 'trunk mains' to distinguish these major pipelines from 'distribution' mains which connect to individual properties. Supply mains frequently connect to a 'service reservoir' which serves a locality.

A modern water supply system is vastly different from when our forefathers were collecting water from a nearby river or lake and carrying it in jars or buckets to the home. Of course, some places in the world still do not have mains water and it is estimated that there are still over 800 million people lacking basic drinking water. The United Nations Sustainable Development Goal 6 (SDG6) aims at provision of safe and affordable drinking water for all by 2030.

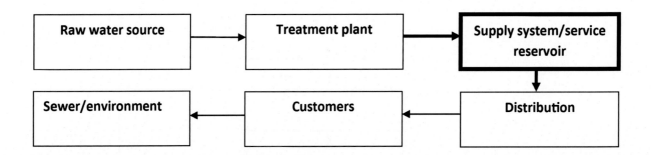

A modern water system starts at the raw water source from where it is transmitted to a treatment plant. After that it goes into the supply system, which may have treated water storage, followed by distribution and then on to the customers' taps.

The supply system requires careful consideration of the topography of the area ; ground levels, existing roads, railways, farmland, other utilities such as electricity, broadband and telephone cables, gas mains and sewers. We also need to consider the demand and the supply must be capable of more than meeting it. Domestic customers require water for cooking, bathing, watering their gardens, washing cars, etc. and this demand generally increases as communities develop and become wealthier. Industrial and commercial organisations often use huge quantities of water; farming, power generation and steel making come to mind. However, industrial processes often re-use much of the water by in-house treatment and recycling. Water companies use historical data to predict future demand in order to ensure that the infrastructure can cope with increases. Systems are also designed to cope with the fluctuations in demand which occur on a daily, that is a 'diurnal' basis. For example water demand generally peaks between 6 and 8 am in the morning and between 6 and 8 pm on weekday evenings. Weekends and public holidays tend to have lower, flatter peaks because there isn't the same get up - and - go to work/come home aspects of a weekday.

An example of the 'diurnal factor' is shown below. The relatively flat line between approximately 2am and 5am is called the 'night line' and operators use this period to assess leakage in the networks. Demand from domestic customers is usually negligible during this period although industries and some commercial properties, such as night clubs, will still be using mains water. There is more about waste, misuse and leakage elsewhere.

Diurnal Factor

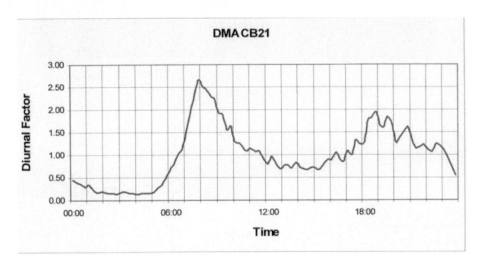

 A typical water tower; this one is in Leicestershire. Note the mobile phone aerials on the top; water towers are ideal for such aerials because of their height and location in urban areas. A water tower will typically hold a reserve of 6-8 hours of water for the area that it serves.

Photo by Alan Sutton

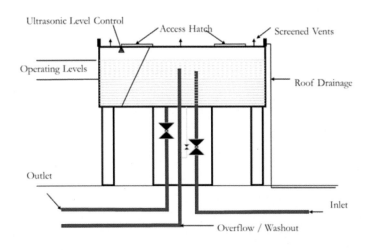

Schematic of a typical water tower showing the pipe work.

Traditionally, instead of an ultrasonic or telemetric control system, towers would have a ball valve which closes when the tank is full, similar to a domestic toilet but much larger.

Photo by Alan Sutton

Hoo Ash Service Reservoir, Leicestershire. This is a typical reinforced concrete service reservoir. The roof is covered with grass in order to blend in with the surrounding countryside.

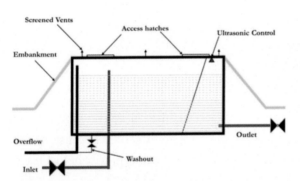

Schematic of a service reservoir

The water level in a tower or reservoir can be controlled either by: a traditional inlet ball valve or a remote pump control. Modern systems use an ultrasonic water level sensor to control the pump.

There is currently a movement away from water towers and service reservoirs in favour of using variable speed pumps to maintain water pressure in the system. Whilst this is now very reliable, it does not cover the hours of supply which are stored in the tank.

Some installations are often quite ornate and grand. This disused tower, near Burton upon Trent is owned by South Staffs Water and now has a role as a radio and mobile phone mast.

Photo by Alan Sutton - Winshill Water Tower, near Burton upon Trent, built in 1907

Zones

A 'water supply zone' is generally an area supplied from a single water source, treatment plant or service reservoir/water tower. A village could be classed as a zone, as could a large out-of-town retail park, business park or industrial complex. The zone is metered and the meter continuously monitored by a telemetry or mobile phone system.

A computer within a control room of the water undertaker analyses the metered flow and if an increase outside established norms occurs the excessive demand instigates an investigation. A specialist team would be looking for:

- Leakage within the zone
- Illegal use (an illegal connection or use of a fire hydrant for unauthorised purposes, etc.)
- Leakage within a customer's area of responsibility

Strategic Grids

Some water companies have a strategic network, within their supply system, which enables them to move treated water from one major treatment plant or reservoir to another in times of stress. The strategic system serving the West Midlands was first promoted during the 1980s and took nearly thirty years to complete. The basis of the grid had already been made when the East Midlands system was connected to Oldbury Reservoir near Nuneaton following the 1976 drought but the connection took over ten years to complete. Following 'The Worcester Incident' two major plants along the River Severn were interconnected and the Birmingham and Coventry systems linked. The grid was completed when South Staff's Bar Beacon Reservoir was connected with a two-way link to Severn Trent's Perry Barr Reservoir in 2010.

The West Midlands Grid

There is a simple principle to learn from this – water companies have, historically, only moved treated water about internally and will not, generally support their neighbours. The converse is – that if we want a national water grid, then it has to involve <u>raw</u> water movement between sources and/or reservoirs (see Appendix 8).

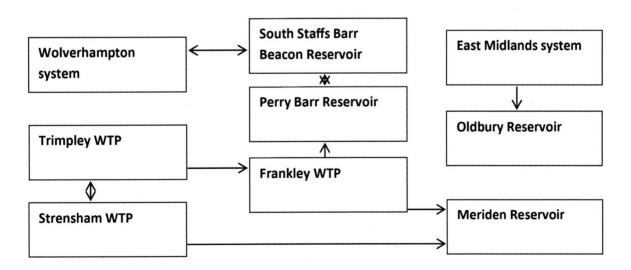

Water Distribution

Overview

The terminology used for the network downstream of the trunk mains and storage system is abbreviated to 'distribution.' It is that part of the network which provides water directly to the customers, who were once called 'consumers' because they consume water without having a choice of supplier. The customers are usually classified as domestic, commercial, medical and industrial.

There is no one single document which sets out what must be done to control, manage and improve the distribution system, it is the experience of the operator to make choices over what are appropriate. However, pipes, fittings, coatings, etc. must all conform to health and safety requirements in addition to material, size and performance standards. Knowledge is passed down and most water utilities will have established codes of practice which are based on years of practical experience. There are a number of sources of reference which can be used and these include: British Standards (BS); European Standards (EN); Water Industry Specifications (WIS); Information Guidance Notes; The Pipe Materials Selection Manual and The Water Supply Byelaws Guide. Together these form a Code of Practice for the UK Water Industry.

Pipe sizes in distribution are smaller than in the supply/trunk main system and in some rural areas they can be as small as 2" diameter (50mm) although 4"(100 mm) and 6" (150 mm) are the most common sizes. The design of the system will vary, depending on the length of mains, topography of the area, especially the variations in the height of customers' properties compared with the height of the stored water or pressure in the supply main.

Open and Closed systems

A traditional distribution network is referred to as an 'open system,' as represented in the diagram below:

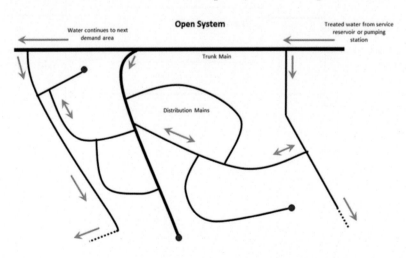

In an 'open system', water flows from the trunk main into the distribution area through multiple branches; there are three in the example above. One problem that arises with an 'open system' is that the direction of flow changes as demand varies throughout the day and season as represented by the double-headed arrows in the figure. Flow reversals disturb any sediment or corrosion debris within the pipes and customers subsequently receive discoloured water through their taps.

Also, an open system does not provide useful data for flow monitoring, leakage control or statistical purposes. This type of network was typical as operators considered it easier to manage. As the need to reduce leakage and to manage the systems better, was recognised, the networks were split into zones which we now call 'district meter areas' (DMAs) as each one is monitored by a district meter.

District Meter Areas

DMAs can be classed as 'closed systems' as the incoming flow is controlled through a single entry point which is monitored by the district meter.

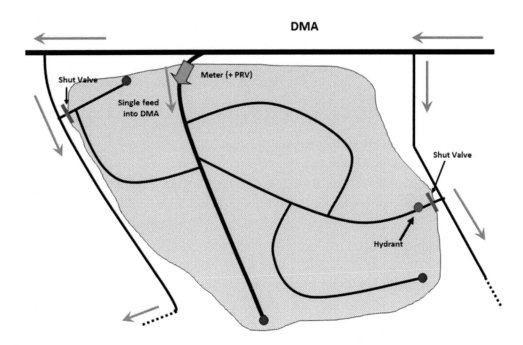

District Meter Area (DMA)

The shaded area is the DMA. It is supplied from the trunk main (or other system) through a single point. If pressures in the trunk main are higher than needed in the DMA then a 'pressure reducing valve' (PRV) is installed in series with the meter. Reducing pressures results in less leakage and hence waste of water. It is good practice to install a hydrant (often called a fire hydrant) at a 'dead end' in the network. Flow rates in a 'dead leg' can be quite low or even zero at times which allows sediment in the water to settle out causing discoloured water problems.

The end hydrants are used by operatives to occasionally flush the system, clearing the network of solid particles and improving the hydraulic characteristics. Hydrants are used for fire-fighting purposes and for monitoring the water pressure in the system by attaching pressure gauges or transducers with data loggers.

District meters are either read weekly by a meter reader or remotely through the mobile phone system so that significant variations in flow can then be investigated. An unusually high reading will suggest leakage although it could also indicate abnormal legitimate or illegal consumption.

Leakage

Water companies are often criticised for not repairing leaks quickly enough, especially when water is seen to be running down a road for several days or even weeks. This lack of response is usually down to economics - if the cost of repairing a leak is greater than the financial value of the water lost - then the company will not rush to carry out the repair.

Ofwat, the water regulator has set water companies performance commitments to reduce leakage by 16% by 2025 from the levels in 2020. If achieved, this commitment will save enough water to meet the needs of everyone in Cardiff, Birmingham, Leeds, Bristol, Sheffield and Liverpool.

In addition to the 2025 targets, UK Water has commited water companies to delivering a 50% reduction in leakage from 2017-18 levels by 2050.

If leakage is indicated by a higher-than-usual meter reading, the approximate location of the leak can often be found by a visual inspection of the area but, for leaks not showing on the surface, operatives will listen for the noise that flowing water makes. Water at high pressure will make a disinctive sound as it flows through the hole or failed joint which is causing the leak. A listening stick can be used by a trained operative or, more modern equipment, such as a 'leak noise correlator' (LNC) may be employed. This device converts the sound of the leak into electrical impulses and the localtion of the leak is indicated by the correlator.

Step-testing can also be used to determine a specific length of water main that contains a leak. A step-test involves gradually isolating sections within a DMA, usually in the early hours, and observing the flow through the DMA meter. A drop in flow, coinciding with the shutting of a branch provides the leakage team with a starting point for further investigation. This would usually comprise turning the water back on and precisely locating the leak with a listening stick or correlator.

It is universally recognised that reducing leakage is an important part of ensuring that there are adequate water supplies in the future. Lower leakage levels will also result in less stress on the environment through a reduction in the volume of water that needs to be abstracted.

Whilst this subject may appear simple in principle, it gets more complex once the finance people get involved. In the UK we tend to talk about 'leakage' and by this we mean that water which is lost from the supply system after treatment and before we get to the customers' system. Looking purely at leakage is an over-simplification of the situation, in the international scene we get the introduction of two more definitions: 'unaccounted-for-water' (UfW) and 'non-revenue-water' (NRW). Whilst many debate the difference between the two terms, NRW includes the financial losses which are incurred by the non-collection of income from the supply. Non-revenue-water is therefore more than leakage alone; it includes water used in fire-fighting, mains flushing, illegal connections and unmetered consumption.

Pressure Control

It is the responsibility of the distribution staff to maintain adequate flows and pressures. Ofwat monitors 'levels of service' against prescribed indicators. The required service level for water pressure was ten, but is now seven, metres head of pressure measured at the external stop tap, and a flow of nine litres per minute at the kitchen tap. This should be sufficient to fill a one-gallon (4.5 litres) container in thirty seconds. Naturally this pressure would be inadequate for multi-storey buildings and in such cases there will exist specific requirements, including possibly a booster pump.

In a relatively flat area, the distribution systems may be supplied at a fairly constant pressure but with varied topography, there is a need to split an area into pressure zones otherwise some sub-areas could have excessive pressure and others insufficient. This is done by splitting the supply areas, as we have discussed above, in DMAs and using valves to isolate them so that a suitable pressure can be maintained. Whilst there are other measures which reduce leakage, pressure control is the most important; the higher the pressure, the more leakage will ensue – but this is common sense – isn't it?

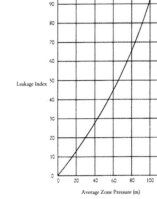

Pressure reducing valves are now used as they are cheap to install and provide greater flexibility than the fixed level of a break pressure tank. The relation between pressure and leakage has been established through research and practical measurements in real distribution systems and can be represented graphically. In general, for every 1% reduction in pressure there is approximately a 1% reduction in leakage.

Ancillaries

As well as the pipes there are many other items of equipment which are necessary to control and manage the distribution system. These include:

- Hydrants
- Tees
- Meters and
- Valves

Valves in particular come in a variety of designs including:

- Sluice valves
- Butterfly valves (usually found near other plant)
- Air valves
- Non return/reflux valves
- Pressure control valves
- Pressure sustaining valves, which are rarely used

Hydrant with sluice valve

Connections

Historically, water utilities have used direct labour for all repairs, new connections, main-laying, etc with contractors reserved for work on larger diameter mains and for construction such as service reservoirs. Now, utilities employ fewer personnel and so use contractors for most activities, working with utility supervisors. Distribution departments are also responsible for new connections for new housing or commercial properties. Every new property requires a metered water supply whereas older, residential properties are often unmetered.

Service Pipe

The pipe going from the distribution main to the customer's property is connected to the main via a ferrule and is called a service pipe and in the UK this is now generally made from 25 mm diameter MDPE which has an internal diameter of 20.4 mm. 32 mm is also used for domestic connections. In special cases, such as a long distance from the distribution main to the property, a larger diameter will be required. In terms of ownership or responsibility the service pipe is considered in two sections:

- The communication pipe carries water between the water main and the boundary of the property. Water companies prefer to have their stop tap at or near this boundary and this will mark the end of pipe work that is the responsibility of the water company to maintain and repair.

- The supply pipe carries water from company pipe work into the property and runs from the boundary to the first water fitting or stop-tap inside the property. The supply pipe, and any water fittings, are the property owner's responsibility to maintain.

If a water meter is fitted this will be near the internal stop tap.

Metering

Metering can be broadly split into two distinct types:

- Bulk metering such as the output from a treatment works or a DMA
- Customer metering

Bulk Metering

Using the old adage that 'you can't improve it if you don't measure it,' companies need to know how much water they treat and put into supply. Although actual flows, quantities and pressures provide valuable information it is generally variations in those parameters that sound alarm bells in the control room indicting that something has gone wrong. 'Water balance' is a technique used to determine what happens to the water produced and, the more components that can be metered, the more realistic will be the data used for prioritising operational and planning activities. A water balance can be represented thus:

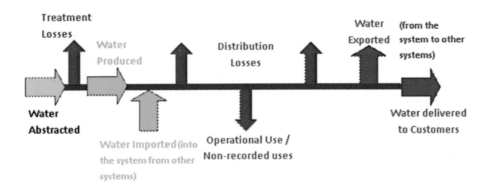

Water Balance

Customer Metering

Commercial properties have been metered since the dawn of time (well not actually quite that far back) but domestic supply metering has been a difficult concept in the UK for political reasons. Many argue that the supply of potable water is a human right and it should not be charged for. This requires an alternative source of funding and local authority rates were generally that source. However, with the setting up of the water authorities, and then privatisation, the need for universal metering has become more apparent. Currently all new housing is metered and existing non-metered customers are given the option of changing over to a meter if they wish. Whilst water meters actually add to the cost of water supply – they need to be tested, installed, maintained and read – they do reduce the demand for water to a small extent, especially in respect of garden watering during dry weather. The introduction of 'smart meters' makes remote reading possible and so reduces the cost somewhat.

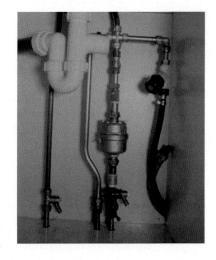

DMAs and Rehabilitation

A District Meter Area effectively provides a tool which aids control and management of the distribution system. It also provides flow data and enables more rapid location of leaks than hitherto and also highlights weak spots, such as low pressures or constant discoloured water problems within the area.

Another tool of network management is computer modelling. This started in the UK in the 1970s but things have moved from the early systems which included: Ginas, Stoner, Watnet and Westnet and now software is based in 'the cloud.' Computer programmes enable engineers to create a model of the distribution network which is then used to investigate the relationship between the network characteristics, pressures and flows as the demand changes minute-by-minute throughout the day. The models are also used to investigate the effects that improvements will have on the network. They can be used to model emergencies such as the loss of a treatment plant or a burst on a major trunk main.

The use of computer modelling, combined with flow and pressure surveys and condition surveys, enables engineers to contribute to the asset management process and investment plans by identifying maintenance, rehabilitation and renewal needs. This ensures that, as equipment and the water networks age, plans are in place to either replace or rehabilitate them. The relationship between pressure, friction and flow in pipes is not precise and there are a number of formulae in use internationally. Hazel-Williams was used until recently almost exclusively by water distribution engineers because it was a much simpler formula and provided a means of relatively easy manual calculation. Whilst the Colebrook-White equation requires a very lengthy mathematical iteration the published tables make its use quite simple. Any theoretical model needs to be verified by means of field surveys measuring flow and pressure at key points in the system. Portable insertion flow meters can be used where traditional water meters are not installed and hydrants used to check pressures.

When actual pressures and/or flows are significantly less than what the model predicts it can be a sign of corrosion or a blockage. If reduced capacity is suspected then a small camera can be inserted into the main via a hydrant to inspect the pipe. However, a more common method is to excavate and remove a short piece of pipe for examination. An example of a corroded iron pipe, showing the products of corrosion, is shown below:

BRIDGNORTH – JUNE 1984
Before Lining

BRIDGNORTH – JUNE 1984
After Lining

Corroded pipe and the effect of re-lining

Coroded pipes, if still structurally sound, can be cleaned and internally lined with cement morter or epoxy resin. Two short sections of main are excavated approximately 100m apart and about a metre of pipe removed from each excavation. A machine is inserted into the water main which scrapes away the products of corrosion, and then another machine re-lines the pipeline. This gives it a renewed life and is less disruptive in terms of causing a nuisance to traffic and pedestrians alike.

Another method, more suitable for heaviliy corroded and weakend pipelines is to insert a plastic pipe (usually uPVC or PE) into the exisiting iron main. The original pipeline is fractured and expanded from inside to enable a larger diameter pipe to be inserted. Re-lining, rather than renewing, negates the need for expensive road and footpath reinstatement - only the excavated portions, where the lining machine has been inserted require reinstating. It remains, however to remake the customers' connections.

Byelaws and Codes of Practice

Similarly to other utilities such as gas and electricity there are rules and regulations to ensure that customers, and others, are not harmed or do not cause harm by undertaking illegal repairs or extensions. In the case of water supply, the responsibilities of plumbers and householders are contained in the Water Supply Byelaws and in essence these laws aim to prevent:

- Waste
- Misuse and
- Contamination

The byelaws are summarised below:

- Customers must not cause waste, misuse, or contamination of water
- Customers must repair or replace pipes and fittings causing or likely to cause waste of water

- Fittings installed in the UK must be approved by The Water Regulations Advisory Scheme (WRAS) whose membership comprises all the UK's water suppliers

- The installation criteria must also be approved by WRAS

- Pipes and fittings must be protected, for example, from frost

- Overflows and cisterns, etc. must comply with the diagrams and rules as set out in the Water Byelaws

- Cross connections between two or more pressurised mains systems are forbidden and back-siphonage must be prevented

Water company inspectors have powers to enter properties where they have reason to believe that the property owner is in contravention of the byelaws. This power is necessary because all the bylaws are designed to conserve water and to avoid any chance of contamination of the water in the distribution system as any contravention could threaten the health of the general public.

Case Study – High Rise Blocks

Water engineers are frequently confronted with choices about how to deal with a problem which often involves topography and gravity. This problem is very simple – how do we get water to the top of a flock of blats (excuse the Spoonerism) which is (say) twelve stories high? Allowing for the need for the header tank to be on the roof, this would need a lift of about 40m which is way beyond the 15m which we are required to supply. The simplest solution, which is to allow a pump to suck off the main, is not allowed in the UK. So, it's sometimes useful to compare systems from abroad and, in this case, we will look at the 'Russian' and the 'Mexican' systems for dealing with supplies to high rise development. They differ in principle as the Russian system supplies the block from mains pressure whilst the system used in Mexico City utilises a 'break tank' with an installed pump to get the water to the roof. In the UK we normally use the 'Mexican' system.

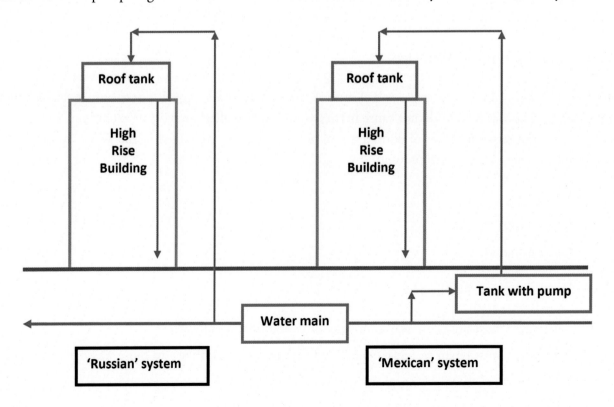

Fire Fighting

Water is the most commonly used substance to fight fires, either using sprinklers, fire extinguishers, stirrup pumps or simply buckets in addition to that used by fire-fighting teams with their fire tenders.

Water companies are not specifically required to lay water mains, nor improve networks, purely for the purpose of fire fighting. However, the Water Act 1991 does include requirements regarding fire-fighting:

- Provide water constantly in mains and other pipes that have fire hydrants fixed to them at a pressure that will reach the top-most storey of every building in the Undertaking's area except in certain circumstances or when work is being undertaken
- Allow any person to take water for fire fighting purposes from any pipe to which a fire hydrant is fixed.

Fire hydrants are fixed on request at convenient points except on trunk mains. The Fire and Rescue Authority, or the person requesting the apparatus, is liable for the costs of this work.

The legislation also states that:

- Water taken for fighting fires or for testing and training is not chargeable
- Recklessly interfering with any resource, water main or other pipe is an offence

Water utilities and fire services work closely together to ensure that hydrants are properly placed and maintained. This is facilitated by a joint group which produced the National Guidance Document on the Provision of Water for Fire Fighting which sets out the responsibilities of water utilities and the fire and rescue services. The document reiterates parts of the Water Act 1991 and sets out to reinforce the commitment by all Water Companies and the Fire and Rescue Service to improve working relationships and secure co-operation in meeting the challenges facing both parties.

Hydrants

Hydrants are generally placed between 100 and 150 metres apart but as close as 30 metres in high risk areas such as where old, highly inflammable buildings exist. They are easily located by the use of yellow marker plates showing the distance of the hydrant from the plate and the diameter of the water main.

The relevant British Standard for the hydrants is BS 750 which specifies a flow of 2 m³/ min which equates to 33 l/s at 17 mhd. In England and Wales the initial minimum flow for fighting a fire is 23 l/s, with rates of over 4,000 l/sec for major fires.

Sprinklers

In the past, sprinkler systems have normally been associated with large commercial or industrial buildings, department stores and factories. Due to recent events, there is growing demand for sprinklers to be installed in all high rise blocks. The provision of water for fire-fighting is on a par with insurance - one hopes it won't be needed but everything needs to be in place for the possibility of fire breaking out. This provision is an essential service, just as important as providing potable water at adequate pressure to domestic customers.

Case Study - High Rise Blocks and Dry Risers

High rise blocks will normally have a large distribution main nearby and this has two separate means of supplying water to the block to fight fires. There will be (as explained elsewhere) a tank and pump which will normally feed the block and this will also supply the sprinkler system if one is installed. A second pipe, called a 'dry riser' is installed in the block with an outlet and fire hose on each floor. When the brigade turn up to fight a fire, they connect the tender pump to an adjacent hydrant and to the dry riser which is then charged with water from the tender at high pressure. Fire fighters then have a ready supply of mains water on each floor of the building.

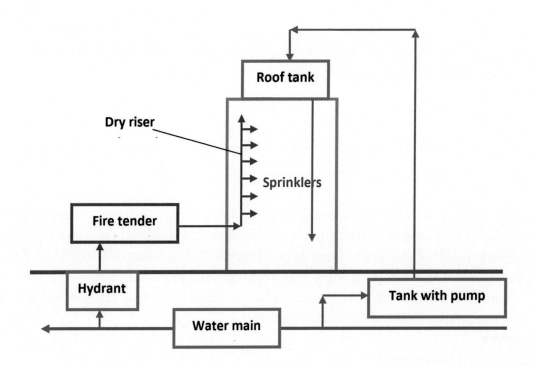

Waste Water (Sewage)

The term 'wastewater' or 'waste water' (take your pick) is common abroad but rarely used in the UK as some confuse it with wasted water which would be leakage. So we call it 'sewage' except for some people who like to call it 'sewerage' as if giving it a posh sound makes it less offensive. The Yanks call it 'sewage.' Either way, it's only something that we have come to live with when we started using potable water in larger and larger quantities and so needed something to take it away. It was many years later that we actually decided to treat it before putting it back into the environment.

As it comes from a variety of sources, it's actually quite a complicated mix of things in solution plus a lot of floating solids and semi-solids. Obviously we all know about faeces, which are a complex organic waste which comes predominantly from humans as your dog and cat are encouraged to use the garden. Interestingly, that's where many of us 'went' in Victorian times and many still do across the world. Urine is a water based solution and contains ammonia which is difficult to treat in our advanced treatment systems but easier in the more natural ones. If you don't want it to create a problem, then just pee on the lawn as this will make your grass grow greener.

Besides flushing the toilet, which will also bring down a lot of paper, much of the flow comes from the bath and hand basin. This is just a weak soap solution with some dirt thrown in but the soaps and detergents that we use contain phosphates which are an essential fertiliser for plant growth. This explains why we get algal growths and deoxygenation in waters that receive such waste without adequate treatment. Washing and cleaning, along with kitchen sink waste produce what we call 'grey water' which is that without faeces and urine.

Depending on the age of your property, the drains may be combined or separate. In the former, the rainwater from the roof and connected paved surfaces joins the foul flows in a single pipeline thus diluting the polluting matter but increasing the rate of flow by a long way. This is why we have a problem with CSOs (combined storm overflows) as the excess flow has to be spilled somewhere if it's not to flood someone's house downstream.

On more modern estates, the foul and surface water flows are kept separate so watercourses are not polluted with diluted sewage. This does not solve the other problem of how to deal with the increased rate of run-off which can cause flooding downstream. The current move to SuDS (Sustainable Urban Drainage Systems) is an initiative that encourages the use of natural retention measures to remove any increase in the risk of flooding.

Some years ago, there was a problem with 'J' clothes as they tended to wrap themselves round the impellor of the local pump. This was obviated by improved design both in respect of the pumps and the clothes. More recently the introduction of wet wipes has added to the occasional problem caused by unthinking persons putting disposable nappies down the toilet. "Disposable" means in the bin – not the loo!

Disposing of fats and cooking oils can cause serious problems – you've probably seen the 'fatbergs' which have recently graced our TV screens.

Alternatives to Sewerage

A 'sewer' is a conduit, usually a pipe used to convey 'sewage' from a property to a place of treatment or point of discharge. 'Sewerage' is the collective noun for a system of connected 'sewers' and is NOT a posh word for 'sewage'. A 'drain' has a number of related meanings and is simple a means by which water drains away. The difference between the 'drain' which drains a property and the sewer to which it connects, is that the drain is usually owned by the property owner whilst the sewer is in the ownership of the company which is responsible. Whilst vested in the operating drainage company, it remains in effect, public property.

The need for sewers only became apparent after properties were supplied with mains water and the occupiers installed flush toilets. Properties in rural areas used a privy for defecation and this was situated in the garden where it would not cause nuisance. It would normally consist of a small shed-like structure with room for one person to sit on a bench with a hole in it through which one disposed of one's personal waste into a bucket which was manually emptied when full. This remains the predominant means of disposal in many parts of the world but is totally unsuited to city life.

When he opened a sewerage conference in Southampton in 1977, the founder and owner of Seer CCTV surveys, Donald Rees told this joke:

"A wealthy American came over to England and fell in love with The Cotswolds. He decided to buy a holiday home there and found a quaint country cottage where an old lady was selling up to move in with her family. After a tour of the place he eventually realised that something was missing and, in a loud voice asked, "But where's the John?" After explaining what he meant, the old lady led him along the garden path to the privy and showed him inside. He was ecstatic at witnessing a piece of living history but then asked with incredulity, "But there's no lock on the door – how does that work?" The old lady thought for a minute and then responded, "Do you know, in all of my seventy years of living here, we have never ever lost a single bucket of shit!"

Victorian slums needed an organised system to dispose of faeces and urine. The latter was sometimes collected and used in business (excuse the pun) processes such as the manufacture of dye fixers. In order to minimise the odours from fresh faeces, they were mixed with the ash from the coal fire and collected by men employed by the local council. This remained the main means of collection until the advent of flush toilets.

Up until the time of WW2, many rural villages had no collection and treatment system. As the properties tended to be disbursed, they simply discharged their sewage into the nearest outlet – the village drain. Whilst this created a severely polluted 'ditch', the problem was usually localised and the outflow from the drain had some natural treatment before getting into the general environment. During the 1950s there was an extensive programme to construct sewers in rural areas, providing localised treatment or pumping to a regional treatment plant.

Whilst over 95% of the UK population has the benefit of a sewer connection, many still enjoy the benefit of a 'septic tank'. The foul drain from the property connects to a tank, or series of tanks, which provide basic settlement and breakdown of the discharged solids. The outflow is then discharged into the environment without further treatment. These cause few problems in areas of low population density and subsoil which absorbs the water. Septic tanks are normally emptied once a year and a residue left behind to enable the natural process to rebuild.

A 'cesspool' is simply a tank which holds discharged sewage until it is emptied, usually by a tanker. It is not designed to provide treatment and simply holds the liquid until it is collected.

In rural areas, it is not uncommon for a property to have its own small treatment plant. If you visit a rural pub, which is situated away for any other properties, you may notice, at the lower extremity of the boundary, that there is a green cover or half dome. This will be their own small treatment plant which could be a three stage septic tank or, more likely nowadays, an RBC (rotating biological contactor). Developed in the 1980s, these small powered plants have a settlement stage followed by series of rotating mesh discs. The natural growth on the mesh provides treatment and aeration before the liquid is discharged. They are sometimes followed by a reed bed.

Overall it is cheaper and often better for the environment to treat things close to their source. Once the population gets denser, or the subsoil is not conducive to absorption, then connection to the public sewer system is the best solution.

Rainwater Harvesting

In the UK we use water butts to save rainwater for garden watering and very little else. However, in many countries, especially where resources are scarce, rainwater harvesting is the norm. Even in Australia, which has sophisticated potable supply systems, some areas rely on rainwater for drinking, washing and cleaning.

Rainwater Harvesting in Gibraltar – 1992 by Jim Linwood
[This artificial catchment has now been replaced by a desalination plant]

Sewerage

The need for sewers

Mankind, in particular and the animal world in general existed largely without the need for sewers. In common with wild animals, men used nature to look after their bodily waste though the results were not always healthy. It's only when you start using large quantities of water for personal hygiene that you perceive a problem. If you live on 15l/d (as many in the world still do) then you don't need a system to manage your liquid waste. However, if you use 135l/d (as most of us in developing countries do) then it soon becomes a problem. Letting it drain into the nearest watercourse creates health problems especially for those downstream and spreading it on land creates its own problems. Thus, the need for sewers to take away our liquid wastes is created by our modern lifestyles and desire for personal hygiene. It is the opposite side of the coin from the production of potable water in large quantities. If we all lived in the forest, instead of towns and cities, and took our water from the local stream, then we would not need them.

What is a 'sewer'?

A dictionary defines a 'sewer' as: "an underground conduit for carrying off drainage water and waste matter" whilst a 'drain' may have all sorts of meanings. The 'drains' leaving a property are the responsibility of the owner and they may carry sewage, and/or rainwater. The sewage will be a mixture of toilet waste water and that from the kitchen which is often called 'grey water' as it does not contain faecal matter or urine. This is further complicated as, in older properties, it may all be carried in a single pipe whilst more modern properties will have 'separate' systems. A 'combined' system mixes foul water and rain water in a single pipe whilst a 'separate' system has two pipes to keep the foul and surface waters apart. Depending on the age of the property, the sewers which the drains connect to, may be separate or combined.

What goes round...comes round

The practical difference between a 'sewer' and a drain was that the drain served a single property and the sewer served more than one. This was brought into law with the 1936 Public Health Act but, from that date onwards, any pipeline which was to become a 'public sewer' required to be 'adopted' by the local authority. Many developers chose not to go through formal adoption procedures and left long lengths of pipeline to be maintained by the property owners under guidance from the local Public Health Inspector. This was rationalised a few years ago and now, all foul pipelines which drain more than one property, are maintained by the responsible water company.

The overall picture is further complicated by the existence of highway drains which are the responsibility of the highway authority. However, before 1974, the sewers and the highway drains were both the responsibility of the local authority so no-one cared about the difference and many were interconnected. This makes the map-based records of who-owns-what crucial when it comes down to asset management and maintenance.

A bit of history

Through the industrial revolution, the development of sewer systems proceeded like topsy and only after 1936 did proper modern systems develop. However, even then, a new estate might have a separate system but the main sewer which it connected to was often combined. This hybridisation meant that overflows were built at crucial points on the drainage systems which allowed the overflow of diluted sewage to the nearest watercourse. Originally called 'storm water overflows' (SWOs), they have become contentious in recent times and are now labelled as combined storm overflows (CSOs). They were generally designed to start overflowing when the flow in the pipeline reached six times the 'dry weather flow' (DWF).

All of this means that every system has degrees of commonality and aspects where it differs from what could be described as normal. Hardly surprising that those who look after sewers, and the designers of new ones, need to understand their history, the law and many other factors.

Egg-shaped sewers

Many considered, as it was part of accepted wisdom, that the egg-shaped sewer was the best shape for a combined sewer. During low flows, the circular bit at the bottom (excuse the technical term) provided the best shape to minimise the friction and the upper part was best for higher flows. The shape was also a little more convenient, than circular, for a man constructing the conduit in a shallow tunnel. They were always made of brick, sometimes with two courses but most frequently with just one and a soft mortar made from lime. Over the years the mortar decayed and a brick fell out causing the sewer to collapse especially under busy city streets.

Brighton Sewer Tour by Dominic's pics

Responsibility and ownership

Once the drains are connected to the public system, the responsibility for the sewer passes to the main water company for that area which does not include the 'water only' companies that proliferate in the south-east of England. This was further complicated as the reorganisation of the water industry in 1974 had a requirement that allowed local authorities to enter into an 'agency agreement' with the then 'water authority'. This allowed them to take responsibility for the maintenance and development of the sewer systems, subject to funding by the overseeing authority. After privatisation, the water companies have gradually wound up these agreements and now have sole charge of the sewer systems.

So, what should happen is that the foul drains connect to foul sewers and surface water drains to surface water sewers. Unfortunately due to their history, surface water often enters the foul system making it combined.

Foul sewers (aka 'sanitary sewers' in some countries)

Completely separate systems are rare in the UK and will only, generally be found in new towns and areas of modern development. They are simplistic to design as the flows are easily calculated based on the properties in the catchment. The size of the pipelines will, therefore, be relatively small and the major problem will be to ensure that the sewage is kept in motion to avoid blockages and septicity. In some areas it was (is?) the practice to connect two gullies at the head of the system to allow some rainwater to enter and flush the system.

The flow is carried directly to the treatment plant before discharge back to the environment.

Combined and Surface Water Sewers

Combined sewers carry all of the sewage and rainwater that flows from properties and often have combined storm overflows which discharge excessive rainwater to the nearest watercourse. Separate systems have two kinds of pipes – foul or sanitary sewers and surface water sewers. Some (hybrid) systems have upstream developments which have separate pipes but they are connected together lower down as there is no adjacent watercourse to connect to.

Whilst not carrying 'sewage', surface water sewers are legally defined as sewers and are owned by the relevant water company. Their design is much more complex as they carry rainwater which is variable depending on the weather. Over the years, design methods have developed to a degree of complexity which now requires a computer to run the calculations. A combined sewer is designed on the same principle as a surface water sewer as the foul flow is small compared with the rainfall carried away. One would expect that a separate system is preferable as it is less likely to pollute the local watercourse. However, the first flush from highway drains and surface water sewers can be very polluting especially after a dry spell.

If you live on an estate and want to find out if you have a combined or separate system, go out onto the road and find a bend. At the bend, there will be only one manhole cover if your system is combined but two, close together, if it's separate.

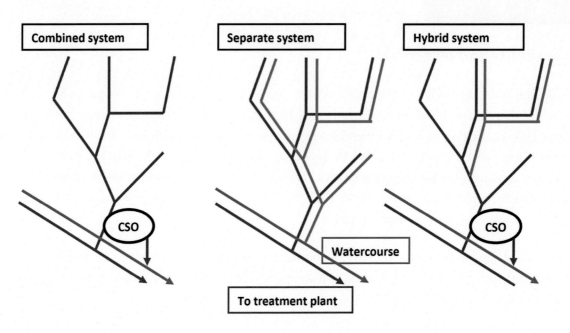

Highway Drains (aka 'storm drains')

The picture is further complicated by the presence of highway drains which are owned by the highway authority. They may be connected to a combined sewer, a surface water sewer or direct to a watercourse. They should not be connected to a foul sewer.

These pipelines carry rain water away from the highway having entered the system via a gully or a channel. The gullies are designed to trap silt and grit and hence prevent the pipeline from silting up and blocking. The gullies need regular maintenance which require the attention of a specialise vehicle – a gully emptier. Whilst they are owned by the relevant highway authority, and often connected directly to the nearest watercourse, many are connected into the nearest combined sewer which means that the water company becomes responsible for the downstream flow. Water companies receive no income for this.

Combined Storm Overflows (CSOs)

Having mixed up the foul flow with that coming from rainfall, we often find that the main sewer is not big enough to accept the increased flow during rainfall. To get over this problem, and avoid properties being flooded by diluted sewage, an overflow is used to spill the excess flow into the nearest watercourse. These overflows were standard practice for many years and some were designed to very basic standards if at all. As many systems were improved over the period 1970 to 2000, the water companies' sewer strategies identified shortcomings and rationalised many systems. In recent years there has been a revived drive to eliminate these overflows and stop their pollution of watercourses.

CSO by Beige Alert

Interceptors

We have already mentioned Bazalgette and The Great Stink in 1850s London. This provides us with the perfect illustration of 'interceptor sewers' as they 'intercepted' minor sewers which outfalled directly into a watercourse. They were, of course, essential in the push to bring about universal sewage treatment. The Thames Tideway project is a new interceptor for London.

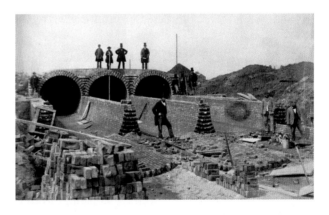

Bazalgette is top right

Sewer Strategies and Drainage Area Studies

Following Severn Trent's development of its sewer strategy, and the publication by WRc of *The Sewer Rehabilitation Manual*, they set out on a ten year programme of *Drainage Area Studies* which could be seen as a forerunner of asset management for sewers. The whole region was split up into 'drainage areas', each containing about 50-60,000 population – this being the size that computers could handle at that time.

The first stage was to ensure that a complete and accurate set of sewer records was available and a basic programme for this set up. A sample of the sewer system was then surveyed using CCTV and the pipelines graded 1 to 5 for their condition. Alongside this, a computer model of the system was built and used to analyse the hydraulic performance using WRc's WASSP software; the pipes were again graded 1 to 5 but, this time, for their capacity to handle the required flows. This information was compiled in an illustrated report along with comments on the performance of CSOs and any other matters such a mining subsidence. An indicative list of projects, which were required to get the area up to standard was then appended and fed into the capital investment programme.

Sea Outfalls

Right up to the turn of the last century, many coastal towns in Britain relied on sea outfalls to dispose of untreated sewage. Many invested in long-sea outfalls and macerators as sea water provides a natural means of treating sewage. Due largely to adverse conditions elsewhere, especially in the Mediterranean, the European Union passed legislation requiring all coastal settlements to have sewage treatment. This was first defined as 'primary' but then upgraded to 'secondary treatment'. In England, Blackpool and Brighton were amongst the last to comply. The introduction of 'Blue Flag' beaches and campaigns by Surfers Against Sewage were also instrumental in bringing in this fundamental change which was largely completed around the year 2000.

Beacons and buoys by Ben Harris

Vacuum Systems

In some very flat areas, gravity sewers are problematical and the sewerage engineer often has to resort to pumping. Some areas, like Florida, use 'lift stations' which are pump stations with no force main but these are rare in the UK. As an alternative, a system using vacuum pumps to suck the sewage along the sewers has been developed. Due to its operation and maintenance problems, it has not proved popular in the UK.

Rodent control

Someone once said that "you are never more than six feet from a rat" but the evidence suggests otherwise as humans outnumber rats by about six to one. However, rats do like sewers and drains as they provide a constant all-year-round temperature with an abundant source of food. Like dogs, rats can eat almost anything and particularly like the waste food which is washed down to the sewers.

Recommended treatments to control them involve test baiting and poison laying. Warfarin was the most popular bait but, due to resistance, more stringent poisons such as fluoroacetamide have been used more often as a rodenticide. Treatments do not, generally, work for long as reinfection soon occurs from adjacent areas.

Leptospirosis, or Weil's disease, is carried by rats in their urine and can be fatal. Sewer workers are required to carry a card with details of the symptoms so that a medical professional will be aware of the risk.

SuDS

Sustainable Urban Drainage Systems have become very newsworthy as many see them as a solution to a number of problems. Wikipedia defines them as: "a collection of water management practices that aim to align modern drainage systems with natural water processes and are part of a larger green infrastructure strategy." In fact, they only deal with the issue of surface water runoff and how to prevent it getting into the foul system or exacerbating flood problems downstream. In practice it means constructing ponds to absorb runoff from development and ameliorate any peak flows.

SuDS near Worcester by Chris Allen

Sewer Safety and Confined Spaces Training

There are two main safety issues associated with sewers. The first involves the safety of pipe layers during construction. This involved regular fatalities until regulations were introduced requiring trench support to be on-site for any depth greater than 1.2m. Current HSE documentation doesn't mention the 1.2 metres rule but states that the need to use trench shoring depends on ground conditions and other risk factors. So the responsibility for site safety is put firmly on the shoulders of all working on, or concerned with, the site and penalties for injuries or fatalities are high. The second is a general concern which is common to many who have 'confined spaces' within their business sphere.

The nature of sewage, as a liquid with biodegradable constituents, means that you can never be certain about what the atmosphere will be like inside a manhole or pipeline. The main risks are:

- Infammables washed down from garage forecourts or other like sources
- Oxygen deficiency due to the decomposition of organic material
- Carbon monoxide from diesel engines
- Hydrogen sulphide produced from decomposition

Prior to the development of modern gas detectors, the industry used a version of the miners' safety lamp with a moist paper which turned black in the presence of sulphides. As technology advanced, each of the risks was

incorporated into the design of gas detectors which are now hand-held devices.

Photo by Mr Zanja

All sewer workers are required to undergo extensive training in the special risks which they encounter when working in sewers. Whilst extensively developed in the 1970s, by the water authorities, and taken up by Water Training, the courses now tend to be contracted out.

Mining Subsidence

Any buried pipeline suffers when subsidence takes place due to its movement relative to the surrounding ground. Up to the 1990s, when most of the UK's coal mines closed, millions of pounds worth of damage were caused each year especially to sewers. Mining typically takes out 2m of coal as the seam is mined and this is reflected on the surface by a subsidence wave which follows the extraction of the coal. It's the relative movement of the wave which creates the damage rather than the amount of subsidence. Pipe line joints are the weakest point and suffer the most.

Whilst there were systems for the recovery of the cost of damage, the burden of proof was such that the system was virtually unworkable. The problem was only solved when the water companies threatened high court action to prevent the privatisation of British Coal.

J Clothes, Wet Wipes and Fatbergs

We all watch on TV, the problems that occur in sewers when things that shouldn't be flushed, are. Years ago, when J clothes were first introduced, there was a problem with small rural pump stations as the clothes wound round the impellor and stopped the pump from working. The problem was only solved by adapting the impellor design.

Currently we have a problem with wet wipes which are not biodegradable and cause blockages when they snag on pipe line joints.

We have referred to 'fatbergs' in our chapter on trade effluents.

Fatberg manhole by Matt from London

Sewage Treatment

Like most industrial processes, the treatment of sewage, or waste water as it is often known, comes in stages. It arrives as a smelly, unsightly liquid with nasty stuff floating in it. So, how do we get it into a state suitable for treatment and then take it on to make it suitable for discharge to the environment? Much of what is still done is based on the 'Royal Commission Standard' which is very simple. It states that the effluent should reach a 20/30 standard which is 20mg/l BOD and 30mg/l SS. BOD is Biochemical Oxygen Demand and SS stands for Suspended Solids. This basic standard has been used across the world but now, more stringent standards are often used especially relating to nitrates and phosphates.

Some configurations will be different, for instance the inlet works overflow may come after screening (which is much better). Many STPs do not have tertiary treatment.

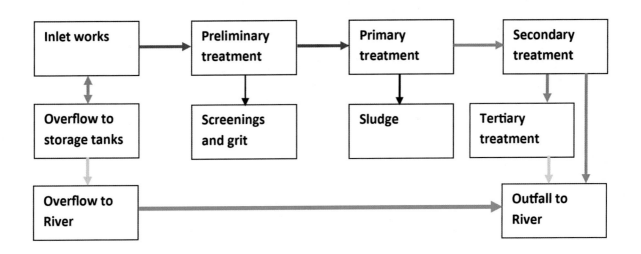

Inlet and Screenings

On its arrival, via the inlet sewer, we first need to make sure that we can process it so it goes through screens. Coarse screens take out the larger items that float down the sewers and then it passes through fine screens which remove rag, paper and sanitary products.

Screenings are disposed of to tip or incinerated.

Overflows

During rainfall the flow will increase and this can cause a problem as the plant will normally be designed to treat three times the dry weather flow (3xDWF). Once the incoming flow exceeds this value then the excess is passed via an overflow weir to a storage tank where it is held until it can be returned after the rain has ceased. If the incoming flow exceeds 6xDWF then another weir comes into operation and the diluted sewage is passed direct to the receiving watercourse without further treatment. Most plants are equipped with screens on the overflow to avoid gross solids getting into the watercourse.

Preliminary Treatment

After this initial stage, there is still a lot of floating material so we pass it through a 'detritor' or a 'grit channel'. These are quite different but both designed for the same purpose, which is - to settle out most of the grit which would interfere with the downstream processes. A detritor moves the liquid around in a tank and allows the heavier grit to settle whilst keeping lighter material in suspension. A grit channel performs the same function by controlling the rate of flow.

Years ago, before we all moved to using tea bags, many plants had a problem with tea leaves which were too light to settle out but too heavy to overflow. This meant that they were trapped in the detritor and gummed up the process. A revision to the design solved the problem but then not many of us wash tea leaves down the drain any more.

Primary Settlement

Our cleaned up sewage is then passed to a set of tanks which allow the majority of the suspended solids to settle out. These 'primary settlement' tanks are normally circular which allows for a scraping device to be moved around the tank. This serves two purposes. At the surface of the water, you can see a device which moves any floating scum to a chute which takes it away. But the main purpose of the rotating scraper is to move the settled 'sludge' to a collector at the bottom of the tank and allow it to be removed by gravity or by pumping to the sludge treatment process. There are many proprietary designs for these tanks including flat-bottomed and hopper-bottomed. The latter are generally more efficient at settling sludge but more expensive to construct.

Aeration

The liquid is now considerably cleaner but still lacks oxygen. So the next process is to aerate it so that it no longer has an 'oxygen demand'.

Historically, after simple land treatment was phased out, this was achieved by 'trickling filters', also known as 'bacteria beds'. They can be circular or rectangular depending on the designer's preference as rectangular beds take up less land than circular ones.

When sprayed or spread onto a bed of stone, the sewage comes into contact with a variety of organisms which grow on the stone, feeding on the impurities in the sewage. As the organisms, most of which are not actually bacteria but small plants and animals, treat the liquid, they add oxygen. As they grow they create their own environment within the bed and different organisms thrive at different levels within the bed. When they die, they produce a waste which then has to be settled out at the next stage of treatment – 'secondary settlement'.

The activated sludge process was discovered in 1913 in the United Kingdom by two engineers, Edward Ardern and WT Lockett, who were conducting research for the Manchester Corporation Rivers Department at Davyhulme Sewage Works. Unlike the tricking filters, which usually operate under gravity, the 'activated sludge' process requires pumping. The settled sewage is introduced into a rectangular tank as a steady stream and it is mixed with a quantity of sludge which is returned for the settlement tanks which follow. This returned sludge has already been activated by the oxygenating organisms which live in it and speeds up the process of aeration in the sewage which is being treated. It is further enhanced by the addition of air which is pumped into the tank from below using compressors, or from above using rotating aerators. Both of these methods are power hungry but, overall, the process is very much more efficient than filters and takes up much less land.

Secondary Settlement

After the ASP, the aerated sewage is passed through 'secondary settlement' where the 'activated' sludge is settled and roughly half is returned to the previous stage of treatment. The clarified sewage should then be suitable

for discharge as treated effluent to the environment depending on the level of treatment required to meet the standards stated in the discharge consent. A basic consent will have standards relating to suspended solids (SS) and biochemical oxygen demand (BOD) which will stated in mg/l. These are based on the impact that the effluent will have on the receiving watercourse or body of water.

Tertiary Treatment

In recent times, there has been a growing need to set consent levels relating to ammonia which is present in sewage from our pee. As it breaks down it combines with oxygen to become less toxic but the resulting nitrogen acts as a basic plant food. In order to grow, plants also require phosphates and these are present in sewage as most washing powders contain them.

Neither of these 'chemicals' are taken out by conventional sewage treatment processes so additional processes are needed. An adaptation of the ASP, using an 'anoxic zone' at the start of the process can reduce the ammonia and hence nitrate levels but phosphates still tend to get through. Sweden, which discharges all of its effluent into the Baltic Sea, just has consent standards for nitrates and phosphates as, if these are within limits, then the rest of the issues will be OK. Tertiary treatment may also make use of micro-filters though these will do little for nitrates and phosphates.

There is nothing particularly complicated in the basic treatment processes of sewage as it's mostly a separation and acceleration of natural processes. If phosphates are a problem (as they are responsible for algal blooms) then banning them from use in the catchment is probably the most efficient solution. Dealing with the levels of dye, perfume and cosmetics can require individually designed, advanced processes.

Nitrifying filter at Wanlip

The Outfall

All STPs need to be located close to a watercourse which is why they tend to be downstream of the community which they serve and adjacent to a river or stream. Obviously, the larger the watercourse, the better the dilution factor and a ratio of one to ten is normally aimed for. Where this cannot be met, the discharge consent may be more stringent than the common standard in order to protect the wildlife in the stream.

Proprietary Systems of Treatment

There are a number of proprietary designed systems that combine the separate processes described above which include those referred to below.

The Imhoff Tank

This consists of an upper sedimentation chamber where collected solids slide down an inclined slope to a lower chamber in which the sludge is collected and digested. The lower chamber requires separate biogas vents and pipes for the removal of digested sludge after about 6 months. The tank is effectively a two-story septic tank which eliminates many of its drawbacks which result from the mixing of fresh sewage and septic sludge in the same chamber.

Photo by Sustainable Solutions

The Pasveer Ditch

This is a proprietary design which could be described as 'an all-in-one' process. After screening, the sewage is introduced into an elongated tank which has a channel running all the way round. The sewage is made to flow around the tank by two large rotating brushes on the surface. These provide aeration as well as keeping the sewage moving. The sewage is retained within the tank for a number of hours before overflowing via a weir and then settling in another tank before release to the environment. This process is quite common in hotter climates but is unsuited to areas which are served by combined sewers as the process is not designed to treat rainwater.

Aeration Tanks

There are a number of other systems which are generally suited to treating low flows and which use a variety of settlement and aeration techniques. They are not common in the UK.

Reed Beds

The reed bed came to the fore in the 1980s and can be used to 'polish' an effluent or for small scale treatment, say, to replace a village drain. When used to polish a 'final' effluent, it is constructed as a wetland which is populated with reeds – usually fragmites Australis. The rhizomes in the roots of the reeds are particularly good at removing trace elements, even heavy metals. When a bed is installed in place of a village drain, or to treat small quantities of sewage, it will normally have a settlement tank or ditch associated with it.

Reed bed at back of treatment plant by Sustainable sanitation

Natural Treatments and Wetlands

The main reason why we use complex processes is to minimise the footprint of the plant. If land is not a problem, then natural processes can be used. Until recently, the City of Melbourne (population four million) used an extensive series of lagoons for all of its sewage treatment. Only when the sludge treatment became too much was a more complex process introduced.

Lagoons

Lagoons are a semi-natural form of treatment which are favoured where land is not a problem. As land is scarce in the UK they are rarely used here but it's still worth examining how they work. As with our 'organised' modes of treatment they work in stages. Screens may or may not be installed at the inlet and they are followed by a primary lagoon which settles out most floating and suspended matter. The second lagoon has a large area to allow for natural aeration. Where the surface area is insufficient, mechanical aeration systems can be installed. The aeration stage is followed by a further stage of settlement which is sometimes referred to as 'polishing'. Each stage, but especially the first lagoon, is normally duplicated to allow one of them to be taken out of action for maintenance.

Sewage Farms

Prior to the early 1900s, most sewage treatment involved the distribution of sewage on to agricultural land which were used to grow crops. They varied enormously in their degree of sophistication and the last one in the Midlands to be replaced with modern treatment was still in use in 1970. The land was simply irrigated with the sewage and it was left to natural processes to break it down and absorb the water into the land or allow it to run off after spreading. Some worked quite well but others didn't and the growing of crops on such irrigated land was eventually outlawed.

Disinfection/Chlorination

Americans, in particular, favour the chlorination of treated effluent before discharge back into the environment. Many consider this to be a complete waste of time and effort for two reasons: (1) if the treatment process is satisfactory then it's unnecessary and (2) it can result in harmful chemicals being discharged into the receiving waters. Take your pick but it's very rarely used in the UK.

Case Study – Kingsbury Lakes

The shallow lakes at Kingsbury were formed by gravel extraction in the flood plain of the River Tame in the 1960s. On west side of the A4097, the lakes are part of Kingsbury Water Park which is owned and run by Warwickshire County Council. On the east side of the road, the lakes have been engineered to form a natural water purification system which treats the whole river, and hence Minworth effluent, to a higher standard.

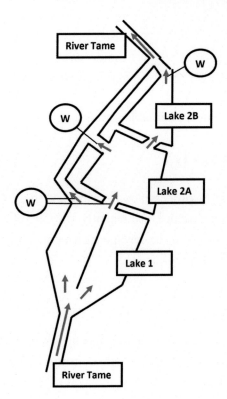

There are four weirs in the system which control the level in the lakes and each provides aeration to the flow. The lakes provide oxygenation via the surface and settlement via the lake bed. There is a trash collector at the inlet to the lakes but it is not currently working which results in the accumulation of a large volume of plastic waste which comes down from Birmingham in times of flood.

Sludge Treatment and Disposal

A problem that almost all industries face is "what do we do with our waste?" and the water industry is no exception. The two biggest problems involve construction waste and waste sludges from the treatment processes. Sludge produced from the treatment of potable water is very thin which creates its own problem. However, if the treatment plant has access to the sewer system, then the obvious solution is to simply dispose of it there. The residual chemicals actually aid the settlement process at the sewage treatment plant.

You will recall that when the sewage was first settled, that the resulting sludge was drawn off at the bottom of the tank. Its subsequent treatment and disposal tends to cause the most problems associated with sewage treatment. Sludge is a little like eggs; there are many ways to cook it/them and when they go wrong they do it for the same reason – sulphides.

Collection from Primary Tanks

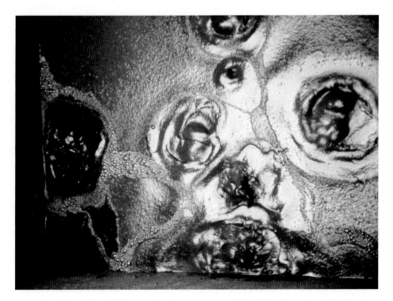

Primary settlement tanks come in two basic forms. They are usually flat-bottomed or hopper-bottomed. The latter are more expensive to construct but are more efficient and generally produce a thicker sludge. The thickness of sludges is a major factor along with the trace elements which settle out from the sewage. How we deal with the sludge and then dispose of it depends, to a large extent, on the contaminants which remain even though they are at very low concentrations.

We have written elsewhere on the influence of trade effluents on treatment but their presence is also of great importance when dealing with the waste sludges. Heavy metals have to be measured and recorded, especially in areas where engineering and associated industries are, or have been, prevalent. The concentration of lead is often used as an indicator and much of this comes from domestic sewage as its compounds form the base for many cosmetics.

Collection from the primary tanks can be done by gravity but this requires that the topography of the treatment plant allows for the required head. Pumping is the usual option and this will require special pumps which are able to handle the sludge which will still contain a proportion of grit which tends to wear the impellors. Whilst centrifugal pumps are preferred, thick sludges may require the use of positive displacement pumps.

Thickening

After removal from the primary tanks, the sludge is typically very weak – say 3-5% solids in suspension – so the next stage is to make it thicker. There are many ways to do this, involving patented designs - of which the picket-fence thickener is typical. This device has a stirrer which is placed in a tank and it stirs the sludge gently. Water separates out and the sludge can get to say 7% which is considered comparatively thick.

Some processes use lime as a thickening agent which also corrects any pH imbalance.

Drying beds

The tradition method to deal with unthickened sludge would have been to simply spread it on land but this was banned many years ago and sludge drying beds were introduced into the process. The beds are constructed to allow the liquid to filter through sand/gravel or some other suitable material whilst the solids remain on the surface. The filtrate is collected at the base of the bed and returned to the works inlet. This works well in dry weather but does not so well when it rains. It's also a very smelly operation and was the commonest source of complaint near sewage treatment plants.

Composting

Composting is a natural process which involves mixing the sludge with organic material and allowing it to decompose in the atmosphere, i.e. in the presence of oxygen. It can be complicated by the need for mixing and turning. It can also create a smell nuisance.

Digestion

Whilst sludge digestion has been used on larger plants for many years, it came to the fore with the creation of the water authorities in the 1970s. Designs and processes were standardised with a commonly used 30/30 target. This implied keeping the sludge at 30 deg C for 30 days which resulted in it being effectively sterilised and less smelly.

Most modern digesters are equipped with measures to draw off the resulting gas which is predominantly methane. The gas is then fed to generators and used to generate electricity. The engines are very similar to marine diesels and can suffer from accelerated wear of the pistons due to the high levels of sulphides present in the gas.

One advantage of having digesters on-site is that strong trade wastes can be accepted on a commercial basis thus spreading the cost of the process. Weaker wastes can be added to the treatment process at the works inlet.

Land disposal

Liquid digested sludge can be disposed of to land, dependent upon the concentration of heavy metals. If the level is very high then the disposal is further complicated. Many authorities disposed of such sludge to 'sacrificial land' which meant that the sludge was ploughed in to fields that were dedicated for the purpose which had any agricultural product carefully monitored.

Disposal of liquid sludges, as fertiliser, is carefully monitored and undertaken using specially designed takers and ploughs using tines which place the sludge at the root level of the crops. The spreading of dried sludge, which appears solid, is undertaken by spreading machines to achieve the required rate of spread.

Incineration

Industrial areas such as Birmingham and the Black Country produce a sludge which has, historically been contaminated with high levels of heavy metals. This means that disposal to land is not possible. The sludge is first digested, then thickened and finally incinerated. The ash, which contains the concentrated residual of the metals, is then disposed of at a special waste tip. Overall, this process is very expensive and hence avoided in favour of cheaper options.

Sludge Strategy

Whilst the supply of potable water is always the prime concern of an area manager, the disposal of sludges is the area which commonly creates the most problems. Most urban areas have always had a clear route for sludge disposal but many rural areas did not. In order to deal with this problematic issue, most water companies have devised sludge disposal strategies. These involve developing one plant as a major reception site with digestion at its core. Following any pre-treatment, minor works from the surrounding area send their sludge to this site for digestion and disposal, usually to agricultural land.

In February 2022, the EA issued a Regulatory Position Statement (RPS231) which updates previous statements about the requirements for the disposal of sewage sludges and the Sludge Use in Agriculture Regulations (SUiAR).

Pumping

The need to pressurise water is so fundamental to the water industry that I decided to devote a whole chapter to it. If you are not into the technology, then you can decide to skip it.

The Need for Pressure

Potable water systems work on very different principles from sewage systems which use gravity wherever possible and avoid pumping as much as they can. Delivering potable water under pressure is a fundamental in water supply and distribution for several reasons:

(1) the water needs to reach the upper floors of a building in order to feed the supply tank

(2) only if the water is kept secure from the atmosphere can you ensure that its quality is maintained

(3) fire fighting can create a severe demand on the system and pressure is needed to ensure sufficient supply.

We sustain pressure in the distribution system in one of two ways:

(1) Pumping to a 'service reservoir' which gives us storage and maintains an almost constant head

(2) the use of direct pumping which now usually involves the use of variable speed pumps which maintain a constant pressure.

Pumping sewage is avoided wherever we can so we only use it where gravity is not available to us. It is often needed where the outfall sewer meets the inlet to a treatment plant as the plant tends to be located in the flat area, downstream from the settlement, and next to the receiving watercourse. Pumping is also common in rural areas where there is always a decision to be made – as to whether to have localised treatment or to pump to a centralised treatment plant. The major improvements to rural sewerage, carried out in the 1950s and 60s, usually involved the latter solution so each and every village has its own sewage pump station.

How pumps work

There are two main kinds:

(1) positive displacement pumps
These work by pushing a liquid much like the pistons in a car engine and so their action is a reciprocating one. They are most suited to pump smaller quantities of liquid and especially ones which are very viscous or those that contain gross solids.

(2) centrifugal pumps
These work by spinning an impeller inside of a casing and so force the liquid out using centrifugal force (purists dispute whether it's centrifugal force or lack of centripetal but let's not get into that).

We sometimes use positive displacement for pumping thick sludges but for our boreholes, raw and potable water and sewage, we use centrifugal pumps to get stuff where we want it to be. There is another fundamental difference between clean and dirty water pumps. In order to achieve a high degree of efficiency, clean water pumps have a narrow outlet from the impellor. However, sewage is not so consistent and so the impellor has to be designed to pass solids – even as big as a cricket ball. Whilst this reduces efficiency, it is necessary to ensure that the pumps do not suffer regular blockages.

Priming

Whilst some designs of pump are 'self priming' most are not and so they require to be 'primed'. This means that they require a liquid, rather than just air, to be in the chamber which contains the impeller. Whilst this is not a problem when we are dealing with a relatively clean liquid like raw or potable water, this is not the case for sewage. Whilst priming is possible for sewage pumps, it is difficult to maintain and operate. Thus the body of sewage pumps is (virtually) always below the level of the liquid level in the wet well.

This explains the very different designs of pumps stations for clean and dirty water. Clean water pumps can be situated at a level which makes them easy to access and hence maintain and are 'horizontal centrifugals'. They tend to be installed alongside the motors which drive them. In the traditional design the sewage pump and its motor are at quite different levels and hence 'vertical centrifugals'. The pump is located below the level of the incoming sewage and the motor is vertically above, usually at ground level. They are connected by a long metal shaft which is driven by the electric motor above.

Wet and Dry Wells

Whilst the priming issue has been solved, we still have issues with operation and maintenance. So the standard design for a sewage pumping station will normally include a dry well and a wet well. The incoming sewage enters the wet well which is separated from the dry well by a wall. A valved pipe between the two enables the sewage to pass through into the body of the pump and this is connected to a rising (or force) main which takes the sewage away under pressure. The main reasoning behind this setup is to enable the pump maintenance to be carried out safely and in the dry.

Submersibles

There are two main applications of the 'submersible', which is a pump with a close-coupled motor, both of which are immersed in the water that they pump. The need for an efficient way to abstract potable water direct from a borehole led to the development of specialised borehole pumps which have multiple stages built into them according to the head that they have to achieve.

In the 1950s, a Swedish professor patented a new pump design which had the motor close-coupled to the pump body. This invention was supported by the design of the seal between the pump and the motor enabling the motor to function whilst submerged. The vast majority of small sewage pump stations are submersibles.

Force/rising mains

In the UK we refer to the pressurised outgoing pipeline as a 'rising main' though, internationally, the preferred expression is 'force main'. I prefer the latter.

Sewage force mains generally rise from the pump station though they may sometimes descend if the pipeline is passing through a valley. If the vertical profile is not a simple one, then an air valve is required at the high point in order to allow air and any build-up of gasses to escape. A wash-out may also be required at the low point.

Septicity

Force mains which contain sewage for a long 'retention time' can suffer from septicity. Because the sewage has no access to a free surface, the natural processes absorb all of the oxygen and the sewage becomes 'septic'. It turns black and the smell increases due to the development of sulphides which are broken down and can be released as hydrogen sulphide – the smell of rotten eggs.

If sulphides are generated and then released, this can cause serious problems in the receiving manhole and pipeline where concrete and steel can be attacked. Measures to combat the problem involve treating the sewage with chemicals or providing a source of oxygen into the force main.

Archimedes

Archimedean screw pumps are commonly used at the inlet to a sewage treatment plant as they are very good at dealing with variable flows and gross solids.

Lift stations

In very flat areas the pump stations may be constructed without any force main. Thus the pumps simply lift the sewage to near ground level where it gravitates to the next station in the chain.

Vacuum systems

These patented systems were also designed for very flat areas and use vacuum pressure instead of pumps. Due to their complexity and need for maintenance, they have not proved popular in the UK.

Surge suppression

With the introduction of plastic pipes, another problem came to the fore in the 1970s. Whilst even the lower, class A pipes were adequate to deal with the required pressure, they were never intended to deal with negative pressures. When a pump shuts down, the backflow prevention valve shuts very quickly. As there is a large volume of incompressible water travelling along the delivery main, this creates a negative pressure at the face of the valve which reverberates in the same manner as 'water hammer.'

If the pressure goes negative then it can cause this lower class of pipe to fail by implosion. Most cases of failure were found to occur where the pipe had been temporarily supported during construction – hence the need for proper pipe bedding.

Trade Effluent

Britain was at the forefront of the Industrial Revolution and this meant that there was a great demand for clean water. Prior to the development of municipal systems, many businesses constructed their own and the industrialised counties still display large numbers of reservoirs which were constructed to serve the mills.

During early Victorian times, when there was little in the way of sewage collection and treatment, many (if not most) industries simply discharged their waste water into the nearest watercourse, often with no regard for the downstream users. Despite legislation, even into the 1990s, many watercourses were grossly polluted with liquid industrial waste as politicians were often reluctant to insist on compliance which might mean a loss of jobs in the area.

Cities and their Trades

Different trades create different effluent and these are often difficult to deal with as the effluent often contains trace elements which are not easily treatable. There has been a general policy in the UK, to mix these industrial effluents with domestic waste as it often (but not always) makes it easier to treat, dilution being a key factor.

Engineering usually involves the processing of metals which require cleaning before going on to be melted, forged, milled, shaped, bent and lots of other things. Water used as a cleaner after these processes becomes contaminated with metal particles and traces of chemicals used in such processes as 'pickling'. Birmingham and The Black Country had most of their liquid industrial wastes collected into the sewer system and taken for treatment but this results in untreated inorganic traces in the sludge which makes it unsuitable for disposal to land. Hence the sludge has to be incinerated.

Kidderminster was famous for its carpets and, whilst it had many carpet firms, there was only one principle producer of the dyed yarn. This firm also made sure that the yarn was mothproofed and their effluent regularly knocked out the aeration process at Oldington Sewage Treatment Plant. A similar business in Leek dyed yarn for weaving and, despite treatment at the STW, the River Churnett changed colour on a daily basis.

When Walkers' Crisps entered into an agreement to have their effluent treated at Wanlip STW, everything went well as their wastewater was largely from washing and peeling potatoes. However the BOD loading on the plant varied widely with the seasons due to the age of the potatoes being processed.

Trade Effluent Control

As you can see from the few examples above, all industrial cities and towns have their own particular problems when it comes to processing their industrial effluent which are subject to a system of control using 'trade effluent consents'. These legal contracts, between the discharger and the water company, contain strict limits on:

- The total quantity per day
- The maximum rate of discharge
- Its temperature
- Its solids content
- Its BOD
- Limits on (usually) inorganic content

The conditions of the consent are strictly controlled by specialists who take regular samples to ensure compliance.

Charges and The Mogden Formula

Payment is exacted according to the consent agreement which is a legal contract. The payment is usually calculated on the basis of a formula – the 'Mogden Formula'. The basis of this formula is to set out a charge based on the flow and the concentration of the effluent which is based on the suspended solids and BOD of the trader's effluent. The charge per m3 is related to the cost of treatment at the plant. There is, however, a serious defect in this formula in the manner in which it is normally applied as it only takes account of the resulting impurities in the sludge in a simplistic manner. If all of the sludge results from the treatment of domestic sewage, then it will normally be suitable for disposal to land whereas, if it contains concentrations of inorganics, such as metals, it may require incineration and disposal to a special waste landfill. This means that that the cost of incinerating the domestic sludge is born by the water company rather than by the polluter.

Oils and grease from restaurants Fatbergs

Fatbergs are a regular problem and result from a partial blockage enabling grease and fat, usually from restaurants, building up in the pipeline. The omni-presence of wet wipes just make things worse.

Decision tree and whether to take it for treatment

Water companies have a choice about whether to accept trade wastes into their system. This is a simplified version of the decision making process.

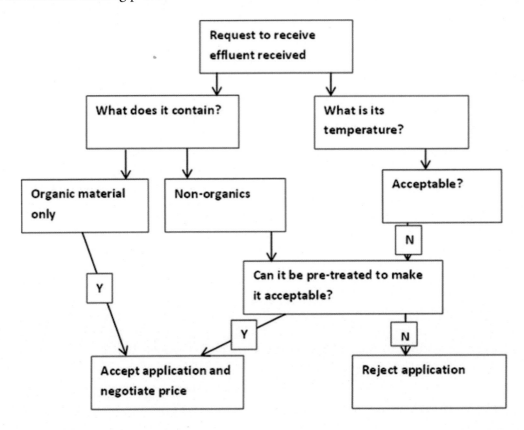

Quality Assurance

Water Quality and Laboratories

QA can be defined as: "the maintenance of a desired level of quality in a service or product, especially by means of attention to every stage of the process of delivery or production." In the water industry this applies to both potable water and the treated effluent which is discharged back to the environment.

QC (Quality Control) is the day-to-day activity which enables the company to meet its quality objectives and thus involves the system operators and those who take and test regular samples to measure what is going on.

Virtually all companies had their own laboratories and even had one on each and every treatment plant. The testing of potable water was always seen as sacrosanct and not to be entrusted to others. However, the automation of standard tests has seen change and laboratory services are now more likely to be contracted out.

DFAT photolab

Very little testing is carried out directly by regulators as the internal procedures of the water companies are trusted to be carried out rigorously and reported truthfully.

Potable Water

The frequency of tests ranges from continuous to infrequent according to the need for the test and the risk that something poses. Since the 'Worcester/Wem Incident' it has become commonplace to automatically test for aromatic hydrocarbons though it was rare prior to the incident. When tap water does not accord with the required standards, the water company may issue a 'boiling notice' which requires any water that is to be consumed to be boiled first. Whilst it was commonplace, this is now quite rarely done.

Listed below are the main tests which are carried out on potable water.

Pathogens

Regular testing for pathogens is required especially where treated water is stored. Sand filters can provide a breeding place for cryptosporidium. If there is residual chlorine in the water, then it is normal to assume that there are no pathogens.

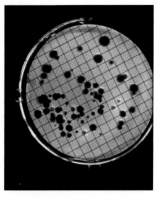

Chlorine Levels

Whilst the concentration of dissolved chlorine may be constant at the exit from a treatment plant, it will gradually decrease as water travels through the system. This needs to be monitored and a 'normal' residual level of 0.1ppm is aimed for.

Fluoride

As fluoride is added by the water company itself, on behalf of the health authority, it needs to be continuously monitored and kept close to 1.0ppm.

Temperature

If water is too warm, it can limit its ability to hold oxygen and decrease its capacity to resist pollutants.

pH

The acidity of the water affects its taste and the corrosion that it causes within pipe networks. It should be as neutral as possible.

Hardness

The concentration of calcium salts is measured to ensure that the water is not too hard.

Chlorides

Chlorides can be an indicator of industrial pollution.

Salinity

A measure of the non-carbonate salts dissolved in water can indicate saline intrusion into shallow coastal ground waters.

Dissolved Oxygen

Clean potable water requires a high concentration of dissolved oxygen as an indicator of its purity.

Turbidity

How clear the water is. If high, then customers will complain especially if washing is affected.

Nitrate and Phosphate

These are an indicator that the source is affected by agricultural fertilizers. The EU issued a directive requiring levels to be reduced.

Pesticides

Traces of pesticides are looked for along with their concentration levels.

Electrical conductivity

The total amount of solids dissolved in the water is measured which also indicates the level of salinity.

Metals

Occasional testing for metals especially heavy metals - aluminium, antimony, arsenic, beryllium, bismuth, copper, cadmium, lead, mercury, nickel, uranium, tin, vanadium and zinc. Heavy metals are known to harm kidneys, liver, nervous system and bone structure. Arsenic is a particular problem with some ground waters.

Sewage Effluent

We treat sewage in order that the effluent that we discharge will have only limited effects on the environment. Over the years we have developed limits and tests which support the enforcement of those limits whether carried out internally or by a regulator.

BOD

The need for an analysis of 'biochemical oxygen demand' is based on what loading of sewage effluent that a watercourse can accept without adverse effects. A river, with multiple discharges along its length, has to self cleanse as it progresses downstream in order for the fish and plant life in it to flourish. Studies by those associated with the Royal Commission, found that, by setting a limit of 20mg/l for BOD, that the majority of rivers would be able to recover by the time the next discharge was made.

The problem with BOD, however, it that it can take five days for the test to be completed and so a simpler COD or 'chemical oxygen demand' is often carried out on site as an indicator.

SS

Suspended solids have also been used for many years, again based on RC requirements. This measures the amount of solid material which is present – suspended – in the water and the RC recommended a limit of 30mg/l. As this is not time or temperature dependent, it can be carried out quite quickly. Depending on the relative size of the material, it is closely related to turbidity.

Finance

Many assume that most of the finance function is just concerned with the accounts but it's much more than that. As they say, "money makes the world go round" and it's just as true for the water business as any other. Without adequate financing, the money dries up followed by the water. Appendix 6 has more detail on regulation.

Treasury

Whilst the term suggests that this part of the business is looking after the gold reserves, they are primarily concerned with the company debt as most businesses split their finances into two distinct parts. The day-to-day spending is called the 'revenue' function and it should be covered by the annual income. But spending on new assets is usually covered by 'capital', that is if the asset is intended to function for more than a year. Capital assets are funded by borrowing which is looked after by the 'treasury'.

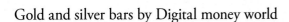

Gold and silver bars by Digital money world

Their secondary function is to make good use of any temporary excess funds which may be available in the bank account. These are lent out to other businesses 'overnight' to cover their shortfalls and avoid bank charges. When our business is short of temporary funds then the process is reversed and we borrow 'overnight'.

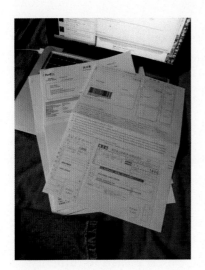

Paying bills here is fun by Kalleboo

Billing and Collection

Historically both domestic potable water and sewerage charges in the UK were covered by local authority rates. Businesses were charged for their potable water according to meter readings and their sewerage charge was included in their rates. Water and sewerage charges are now set by the water companies, which supply the service, but monitored by Ofwat. In areas with a split service it is usually the water supply company which serves a single bill for the entire service. There are three main ways that a water business can be financed though only two have been prevalent in the UK. Abroad, it is not uncommon for the central government to provide a proportion of the financial needs of a utility especially if it is an arm of government. Prior to 1974, the funding of water services was often included in the local authority rates either as a direct charge or as a precept if the service was provided by a water board. This explains the lack of metering for domestic supplies in the UK. After reorganisation, the water authorities made their own charges but this was complicated as they had to cover for potable water supply, sewerage and sewage treatment, plus something to cover surface water drainage.

Metering makes thing simpler though it has its own problems as it involves the cost of meter-reading and maintenance. According to Water UK about 50% of English domestic properties have their water supply metered though all businesses are metered. Since 1990 it is compulsory for all new homes to have a water meter. Even with a well managed system, only about 97% of billed water is collected for. The argument about water being a human right ignores the need for its treatment and pressurised supply and the issue of other losses, including leakage continues to this day.

Accounts

The accounting section, or 'bean counters' as they are affectionately known, keep track of the financial transactions that take place. Every invoice sent out and those received are all 'accounted' for in what we used to call 'book keeping'. Whilst we use computers these days, many still use that term. 'Double-entry book keeping' is a skill in its own right and still forms the basis of accounting.

The section gets very busy during March as they have to tally up every transaction for the year and produce a set of accounts by the end of the financial year.

Accounting by Zolierdos

Management accounting

Management accounting, as a separate function, only came into its own in the 1980s. Its purpose is to set out the financial arrangements for every manager in the business through the use of 'budgets'. This is based on an agreement between finance and each of the spending departments, about how their finances are structured and the amounts that are expected to be incurred during the year. The use of spreadsheets has made the whole exercise more understandable and will, generally by provided in two forms – 'objective' and 'subjective' linked tables. The first says what the purpose of the spending is, and the second one has headings like manpower, transport, power etc.

Payroll inserts by Kevin Dooley

Payroll

The payroll section have to ensure that everyone is correctly remunerated and paid on time.

Procurement

Since the purchase of goods and materials requires that they be paid for, the function is normally part of finance. The Finance Director is also responsible for the signing of contracts which are usually prepared by other departments. If not properly run then the whole business can get into trouble and even collapse due to simple issues like lack of spare parts.

Stores

It's debateable whether the stores function should come under Finance (who procure items) or be part of Operations (who use most of them).

Audit

The kind of audit that is generally understood is the checking of transactions and accounts to see that everything is ship-shape with no discrepancies or illegalities. During the 1970s, a number of the large water authorities started a process of 'efficiency audit' which looked at the detail of a department and its finances. Carried out by a financial auditor accompanied by a technical specialist from another area, they made recommendations for improvement based on good practice and comparisons with similar departments elsewhere.

IT

Information technology, in its early days usually involved the computerisation of the payroll then billing and collection. Those good-old-days involved mainframe computers with punched cards or magnetic tape as input devices. Due to their use in finance, it was normal practice for the function to be managed within the finance department and the manager of the IT section was often paid more than the Finance Director that he reported to.

Mainframe Computer by Naotakem

These days the availability of packages is mind boggling. Many are available on the cloud or will run on PCs. For instance, MIKE+ has a wide range packages which support modelling of distribution systems, collection systems and rivers.

Pensions

Prior to privatisation, pensions were provided through the Water Authorities Superannuation Fund (WASF) which was administered as part of the Local Government Superannuation Scheme (LGSS). Staff made contributions at 6% of their gross salary and other workers at either 3 or 4%. In 1989, the privatised water companies had to establish their own pension schemes and provide a choice between two 'direct benefit' schemes. One was the primary company pension scheme, which new or transferring employees could join whilst the other is usually referred to as a 'mirror image scheme' (MIS).

There are two fundamentally different types of pension provision found in water companies today. Defined Benefit (DB) schemes, where an employee's pension is determined by salary (e.g. final salary or career average) and defined contribution (DC) schemes where pension contributions are invested in funds to achieve a return.

Over time, the private sector have replaced DB pension schemes with DC schemes which has shifted the risk from the employer to their employees. The DB schemes have evolved over time and many have closed altogether. Nevertheless, these schemes remain financially material to water companies, especially where there is a deficit between the liabilities and assets. A 2018 report indicated funding levels between 77% and 109%, although this has improved of late.

The issue of pension provision is further complicated by the introduction of statutory minimum requirements. Other parts of the industry have their own schemes which all have their own features.

Competition

I wondered whether to make this a chapter in its own right but decided not to as:

(1) it doesn't make sense
(2) the closest model – that of the power industry has patently failed and
(3) I don't agree with it

To envisage one company putting water into the pipes of another for delivery to a remote customer doesn't add up and the idea of the reverse situation involving sewage simply beggars belief. We already have a degree of competition in the performance and price comparisons carried out by Ofwat though some of the ideas that they come up with do strain one's understanding. But then if they are there to carry out a job, then they will certainly find a job to do. Enough of that – if you want to know more about competition, try *GB Water Industry for Dummies* by Dr Graham Haimsworth which has more on that subject than it does on sewers or water mains.

Human Resources

Recruitment

The satisfactory provision of water and sewerage services is a combination of having the right infrastructure in place, the local geography/topography, suitable technology, finances and well trained, experienced and skilled staff. The initial recruitment process is therefore designed to ensure that, as much as possible, the selection of people who, along with in-service training, will bring an appropriate work ethic, skills and ultimately experience to the service provision.

Thus, one of the main functions of HR is to ensure that the business has the right numbers and type of staff to ensure that it can run efficiently. Much of this task can be seen as administrative in that it is largely reactive to requests from the functioning departments to recruit staff when a vacancy occurs. This involves first deciding whether a post can be filled internally by promotion or movement and then advertising it internally based on a job description. Whilst this will state the nature of the job, it will include any necessary qualifications.

"Your country needs you" by Halloween HJB

If the post is to be filled externally then the procedure involves advertising, followed by accepting and sorting applications into a short list, prior to holding interviews. Whilst these are normally organised by personnel, their main task is to advise the functional managers who are responsible for selecting the successful candidate. The formal appointment is made by personnel who confirm the contract conditions and salary.

Payslip by DH Wright

Employment

Virtually all businesses will keep personnel files and these are referenced by name and date of birth. Certain data can be kept without the employees' permission. Examples include sex, dates of current and prior employment plus references from previous employers along with qualifications. Other data such as race and ethnicity, religion, health and medical conditions, is classed as 'sensitive' and requires permission for it to be kept from the employee. For any inventory that is held, the employee has a right to know what is held and how that data is used. The Equality Act specifically identifies nine protected characteristics for an individual employee:

- Sex
- Age
- Race
- Disability
- Religion/Beliefs
- Sexual Orientation
- Gender Reassignment
- Marriage/Civil Partnership
- Pregnancy/Materni`ty/Paternity

It is likely that other data such as the employee's sickness record will be recorded as well as their performance reviews. HR will be responsible for overseeing the issue of verbal and written warnings in the case of employee misbehaviour and will add details to the employee's file. In extreme cases they will handle dismissals. In happier circumstances, they will organise an employee's leaving or retirement.

Employees also need to be provided with the necessary accommodation and tools to carry out their functions. Facilities management may be contained within HR or located with other administration functions.

Health and Safety

H&S is often treated with derision in some quarters as there have been many examples of it being taken to extremes with consequent bad publicity in some quarters of the press. This should not underestimate its achievements as many industries have seen their accident record plummet as simple safety measures were introduced.

The water industry had several aspects of its business which needed to be addressed as new H&S legislation was brought in through the 1970s and onwards. Death and injury due to trench collapses were common but regulation made safe practices mandatory. Working in confined spaces posed specific risks and water authorities embarked on extensive training commencing in the mid 1970s. The use of chlorine in gaseous form poses specific risks and has been reduced in favour of liquid forms. Otherwise, the main risks in the water industry relate to working with rotating mechanical plant and in proximity with deep water.

H&S by Leonard Bentley

It is debateable whether the task should be undertaken as part of the HR function or absorbed into Operations where most risks are encountered.

Qualifications

The water industry, like most other businesses, prides itself on its support for those who want to be qualified. It accepts a wide range of certificates which are appropriate for the skill requirements of the jobs at all levels. They could be seen to be based on this model:

- NVQ1 for Operatives/Manual Workers
- NVQ2 for Senior Operatives
- NVQ3 for Supervisors
- City and Guilds for Craftsmen
- Diploma/BTEC for Middle Managers
- Degree for Senior Managers
- Higher degree/MBA for Directors
- PhD plus specialisation for Specialists

Whilst this framework provides a basis, it does not preclude those with appropriate skills and experience from rising to higher levels in the organisation. These are typical for the UK and there is, obviously, a high degree of overlap between the levels. A slightly different version may be found on Wikipedia.

Whilst university degrees are seen as a basic requirement for those at senior management or specialist grades, the membership of a professional institution is often seen as the pre-requisite for appointment to a senior role. The main institutions involved in the water industry include: civil, mechanical and electrical engineers; chemists and biologists plus the over-arching CIWEM. A variety of management, legal and financial qualifications are evident in support functions.

Training

Virtually all businesses depend on the expertise of their staff in order to succeed. Central to this is training and personal development. Experienced staff may be recruited to senior posts but many are brought up 'through the ranks' within the organisation. A programme of personal development is essential if the business is to keep apace of staff turnover and with external demands.

The opposite side of the 'training' curve is 'learning' and this is something that is unique to each person. Whilst training may be delivered in standard formats, learning depends on the models which suit each person. Delivery may be in a variety of modes which could include:

- Lessons (as at school)
- Lectures (delivered by a lecturer)

- Tutorials (delivered by a tutor)
- Book and paper learning (by reading)
- Presentations (PowerPoint for instance)
- Practical hands-on training

As different people learn in different ways we normally use a combination of the above. Some prefer text whilst others like pictures and diagrams. Whilst some like written text others like to hear it spoken. This is why modern methods such as PowerPoint are so powerful. Those who are numerate like to see numbers and tables of data whilst others see confusion – only to see things clearly when the key points are pointed out graphically.

It is important to differentiate between 'education' and 'training'. Education is experienced at school, college, university and whilst watching the news or a documentary on TV. In terms of school, college, etc. the aim is to transfer as much knowledge about the subject as possible, usually, to pass an exam. For example, the pass mark for obtaining a Third Class Honours degree at a British university is 40% whilst a First Class Honours requires 70%.

Schoolroom by Alan Burnett

Training is different from education in that it aims to transfer skills and knowledge to undertake specific tasks and ideally requires 100% pass in any assessment or examination. It is pointless for an operator or technician to only understand, say, 40% of the job in hand and not have the required knowledge and/or skills to undertake the other 60%. We're sure you wouldn't want to be operated on by a surgeon who only got 60% in his/her practical exams, nor be defended by soldiers who only understand 50% of how to use a rifle. In the case of water supply we expect operatives to fully understand the processes that they are responsible for and that managers understand the whole process.

This is a training course which we set up in Coventry on the Sowe Valley Sewer. You've all seen the tunnel from the inside as it's the one which the three Mini Coopers traverse in *The Italian Job*.

Training will generally include not only lectures, presentations and discussion but, where appropriate, practical work. An adage used by training professionals and attributed to a Chinese scholar, Xunzi (340 - 245 BC) is:

- I hear and I forget
- I see and I remember
- I do and I understand

BTEC – A Case Study

The current water industry has gone through many phases which reached their peak with the delivery of the BTEC modules in continuing education. Why they were discontinued remains a mystery to this day and current methods struggle to get back to their standards. Much of this book seeks to redress a small part of that shortcoming by providing a foundation.

The BTEC Certificates in Continuing Education were awarded for attendance and satisfactory performance on a year- long course which mixed weekly, half-day sessions with 'week away' courses. There were five main subject areas:

- Water Treatment
- Water Supply and Distribution
- Sewage Treatment
- Sewerage
- Management

Each of the five courses was split up into about eight modules to cover the main headings and these were sub-divided into units. Each unit was delivered by a 'tutor'; note – not a 'lecturer' and was very interactive with the sixteen students enrolled each year. At the end of each unit, the students were set a mini project to bring for presentation at the next week's session. As an example, in the first session of the Sewerage Course, they were asked to describe the drainage system and waste treatment of their own property. The students were encouraged and given support in report writing and presentation skills to aid their all-round performance.

Water Training (later named Water Training international, WTi)

'Water Training' was a division of The National Water Council and headquartered in Tadley Court, Hampshire. After the 1974 reorganisation, the industry absorbed or set up a number of residential training establishments across Great Britain. They were:

- Tadley Court in Hampshire (management training)
- Millis House near Derby (predominantly for technical skills and for manual workers)
- Burn Hall near York (general water sector skills and laboratory training)
- Melvyn House near Ayr (large diameter mainlaying and general training for Scotland)
- Flint House near London

Tadley Court by David Exworth

Management training concentrated on issues which needed to be improved in senior employees and potential managers. These included financial and communication skills. Manual workers were trained in practical skills such as pipe laying, connection fitting and pump maintenance. Safety training was always at the forefront especially where confined spaces and working at heights are concerned. Specialist training, such as IT, was generally outsourced.

The centres were financed by a per-capita charge on each of the companies (a training levy) and most of the training was provided by senior/experienced employees of the water companies. The training levy ceased in 1984 as an indirect consequence of the dissolution of the National Water Council by the government. WTi continued to provide training until the new companies started to introduce their own internal measures resulting in rather patchy outcomes to say the least. In spite of branching out into other sectors such as gas and overseas projects, WTi finally ceased to trade in the late 1990s although 'Develop' (part of the JTL Group) continues to provide some of the courses previously provided by WTi.

Unions

Whether or not to join a trade union is very much a matter of choice for each individual. A recent report showed that just under 30% of employees in all sectors of the water industry were members of a union.

UNISON supports more than 17,000 members working in water companies, the Environment Agency, passenger transport executives, bus companies, the Canal and River Trust and regional and local airports. GMB have over half a million members, many of whom work in the water industry. They regularly campaign to bring the water companies back into public ownership. In addition to the power industry, Unite has members in a broad range of occupations across the UK water industry.

GMB workers lobby of the Environment Agency by Roger Blackwell

Maintenance

Overview

Operations and maintenance are often overlooked when considering the relative importance of functions in the industry but operators are the people who provide the service to customers - they are the ones who carry out the basic day-to-day maintenance and keep plant running. In fact, the operations staff would claim that they are the <u>most</u> important people in the utility because they have direct responsibility for delivering the services that the customers pay for. Their place in the cycle of tasks is key.

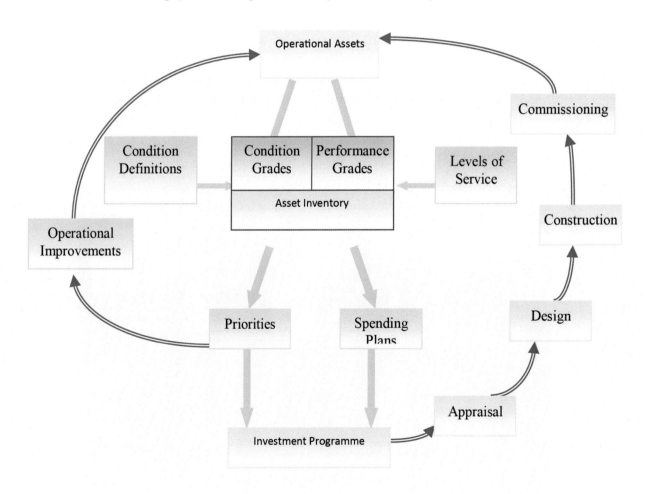

Maintenance Versus Repair

What do we understand by the word 'maintenance'? if you think it's all about repairing things then you would be wrong - the maintenance regime is there to <u>avoid</u> having to make repairs. The OED defines it thus: *'cause to continue; ... preserve or provide for the preservation of (a building, machine, road, etc.) in good repair'*. Repair is defined as *'restore to good condition after damage or wear, renovate or mend by replacing or fixing parts. The act or an instance of restoring to sound condition'*.

In other words, <u>maintain</u> things as the designer and manufacturer recommends and the need to <u>repair</u> them will be minimised.

Corrosion and Degradation of Fixed Assets

Corrosion is the enemy of both underground and above-ground assets. All metallic components, especially iron and steel are liable to corrode in to damp conditions, hence the need to protect metal pipes, nuts and bolts etc. both before and after they are installed. Valve houses, service reservoirs, water towers, pumping stations, treatment plants all have areas that can be subjected to decay in the damp and wet conditions that prevail around our assets. Even inside a reinforced concrete service reservoir there are weak areas where a combination of oxygen and water poses a threat. Ladders, railings, bolts and iron/steel fittings are all subject to this so special attention must be paid to metal components. Rust stains indicate corroding steel reinforcement bars but the main problem with brickwork is the mortar especially when only lime based mortar was used. Cement based mortar lasts much longer.

Tribology in Relation to Moving Assets

Tribology is the science and engineering of interacting surfaces in relative motion. It includes the study and application of the principles of friction, lubrication and wear. Our main interest is in the lubrication of moving parts as this is perhaps the most important factor in keeping rotating and reciprocating plant functional. Without getting into the actual science we need to be aware that manufacturers recommend specific lubricants for a purpose and not just because they have shares in the company. Using the correct lubrication, as recommended in the maintenance manual, is crucial. No-one should economise where lubrication is concerned and I don't just mean sex!

Relationship to Asset Management

The maintenance system deals with 'the-here-and-now' whilst the asset management system covers what we need to do for the future. Asset management identifies the investment needed over a long, rolling period (usually at least five years) in order to keep assets in good operational condition by replacement and rehabilitation and is, therefore, a 'capital' function i.e. funded by borrowing. Sound maintenance ensures prolongs the life of the asset and is, therefore, a 'revenue' function i.e. funded from the current year's income.

There are subtle differences in the needs of the supporting systems. Often over-simplified as the lubrication of rotating mechanical plant, maintenance needs a level of detail which is an order greater than that required for asset management. Whilst the latter generally records assets at the process level, maintenance requires each individual item of plant to be recorded. In addition, it tends to concentrate on mechanical and electrical plant rather than structures.

Reactive and Preventive Maintenance

Most of us have our car serviced at every 10,000 miles or once a year along with more frequent checks on tyres, oil and fluid levels - that's the 'gold standard'. At the other extreme, some of us do nothing until we need to and we suffer the consequences when disaster strikes. A middle of the road (excuse the pun) approach would be an annual service and the occasional check on tyre pressures.

Maintaining water and sewerage systems is, of course, much more complex than maintaining a vehicle but the principles are the same. For one thing most of the assets, i.e. the pipelines and associated ancillaries and fittings, are underground. Therefore it's difficult to have a common strategy for both above-ground equipment such as pumps, treatment plants etc., and buried pipelines.

Management have to decide on its overall approach to maintenance which can be set out as a form of hierarchy, starting with the lowest:

- failure based systems whereby action is taken only after failure has occurred i.e. breakdown maintenance
- minimal maintenance of critical plant whereby regular maintenance is scheduled only for those items of plant which are essential to maintaining the service
- regular maintenance of all plant involving a system with daily/weekly and monthly schedules of tasks for all items (see later note on ISO 9000)
- technology based monitoring and control – such as vibration monitoring to predict the early signs of impending breakdown – and telemetry to provide warning of failure
- predictive systems based on plant history which formalize the instinctive knowledge of operators who generally know which items of plant will fail

Most businesses will choose a system that is somewhere in between these approaches and will have some form of computerized support system to organize and keep track of things (see note on CMMS). The maintenance of linear assets is more problematical. Normal, or day-to-day maintenance usually involves simple actions such as flushing pipelines or jetting to clear a blockage. The repair of bursts and collapses is only done in response to incidents. Most other maintenance, involving rehabilitation etc. would generally be considered to be of a capital nature and therefore part of the AMP and/or a strategy.

CMMS

A planned maintenance system can be paper-based but is more efficient and effective if computerised, i.e. a Computerised Maintenance Management System or CMMS. Some organisations develop their own systems but it is generally more cost-effective to purchase a commercially produced package, modified where necessary to cover the whole range of infrastructure and equipment used by the company. Some packages include aspects of asset management and hence cover the procedures involved in the life cycle of assets i.e. design, construction, commissioning, operations and maintenance plus decommissioning or replacement of the assets.

There are many packages on the market to assist businesses and all have their strengths and weaknesses. Perhaps the best known is IBM's Maximo. Many now work in 'the cloud' meaning that an up-to-date version is always available and data is securely stored.

At the very least the asset management and maintenance management systems should have a common form of referencing so that each of the system users can easily compare information in the two inventories. The setting up of a computerized system requires a similar initial procedure to that employed with the creation of an AMS but goes into much more detail:

- set up a detailed inventory of all plant

- identify critical items of plant

- identify plant with high failure rates

- set up schedules

- set up procedures

- implement and test

- revise and fully implement

All of this seems a little mundane but then there are many bells and whistles which can be adopted, some of which will improve the system and some will not. Amongst these is 'Reliability Centred Maintenance' or RCM which concentrates maintenance resources on those items of plant which are most likely to fail and will often consider monitoring and telemetry to warn of failure.

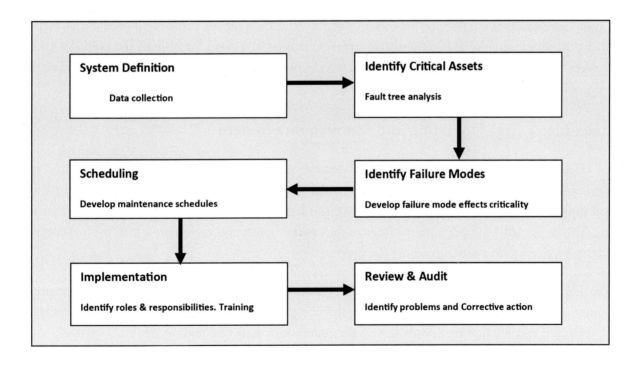

New Technology and its Effect on Operations

'SCADA' stands for Supervisory Control and Data Acquisition, or at least it did the last time I was acquainted with it. The problem with new technology is that it's new and hence everyone and his dog wants to invent a new clever acronym. We have already discussed CMMS or whatever it's called now but there is a lot of good stuff out there.

In the 1980s, the East Worcestershire Waterworks Company set about automating their plant at Sugarbrook and their MD became the industry's guru on the subject. Its big neighbour, Severn Trent waited in the wings while others forged ahead, especially Yorkshire Water who were the first big water authority to invest heavily. The mindset of many seemed to be based on the old Scottish saying "He who pays last, never pays twice."

Whilst some interesting work was done within the authorities themselves, most advances were promoted by the equipment manufacturers who perceived the competitive advantages. Perhaps one of the best examples was the automation of sewage pumping stations. In their early days they had to be manned but once submersible pumps became the norm, the pumping was controlled using probes which measured the depth of liquid in the wet well and turned the pumps on and off. These simple controls were augmented when alarm systems were introduced which monitored the pumps and used telephone lines to warn a central control of problems.

In 1980, one large water authority employed 11,500 personnel and there was always someone present on a water or sewage treatment plant. Today that figure is around 6,000 and many plants are remotely monitored as, in addition to the monitoring, modern plant is much more reliable and hence does not often break down. The numbers are not untypical of the industry as a whole.

We are now in a situation where we look for every opportunity to automate and remotely control. We have computers which can react instantly and more reliably than humans, plus a telecoms networks which is intelligent as well as reliable. Pressure control of distribution networks and the detection of leaks is now commonplace as well as the monitoring of water quality at each stage on a treatment plant. Each step which advances the automation of the networks and plants, reduces the need for human intervention and hence manpower. With this comes the risk that an organisation is unable to respond to emergency situations as they do not have the manpower.

District Meter Area Operation and Maintenance System

Using transducers, data loggers and wireless telemetry systems water companies now monitor residual pressures in real time. For example, a data logger and transmitter are permanently fitted to a hydrant at a critical point in the network. This will usually be the highest point hence the place where the pressure is lowest. The logger will send pressure readings to the central operations computer which in turn controls the relevant pump or valve in order to adjust the pressure.

Whilst this is an excellent use of modern technology, some now go one step further - instead of simply trying to maintain the pressure at, say 15 metres head, the computer will also analyse variations in pressure above this level and if it is fluctuating unduly it will alert operations staff to a possible fault. An investigation will then take place to identify the issue and resolved it before customers become aware of the problem.

M&E Plant, Pumps and Control Equipment

To many in the maintenance profession it's all about rotating mechanical plant and the motors which drive them and CMMS certainly targets this area. All installed M&E plant will come with manufacturer's recommendations, usually in the form of a 'maintenance manual'. This will contain information about: what the plant is intended to do; the correct installation; safety requirements, regular maintenance, including lubrication and a chapter on fault-finding. The manual is often used as a basis for simple instruction charts, which are attached to the plant to cover: start-up, shut-down and problem resolution.

Modern control equipment makes great use of computer chips and bespoke software so specialist advice may be necessary when dealing with problems.

Vibration Monitoring

In the 1980s, technology became available for rotating mechanical plant to monitor whether it was performing satisfactorily. Usually the first sign of impending failure (as operators have always known) is an increase in vibration as the bearings are usually the most susceptible components. Monitoring systems have sensors which pick up vibration and compare the level against the norm. When it exceeds that set, a message is sent to alert the operators who will respond to see what is happening.

Water Mains – flushing, valves

The maintenance of water mains is generally confined to flushing to remove sedimentation which can be caused by corrosion of older iron pipes. This sediment causes discolouration of customers' clothes during the washing cycle and is especially prevalent in older systems and where build-up occurs due to low flows.

Sluice valves need to be slowly closed (or nearly closed) and re-opened occasionally in order to ensure that they are fully functional - should an emergency shut-off be required. Sluice valves, air valves, hydrants, PRVs etc. all need 'working' occasionally to ensure smooth, leak-free and precise operation.

Sewer Maintenance

Sewers are generally left to their own devices as long as there are no apparent problems. Once a problem is suspected then a survey will be conducted which will usually involve 'lamping'. Two manholes are opened and a lamp is positioned in the invert of one whilst an operative looks from the other access point to see the lamp. If a full circle is viewed then the pipeline is OK but, if not, then further investigation is required.

CCTV is regularly used to identify problems in gravity pipelines and a grading system was developed by WRc in the 1970s. Whilst updated, this still provides the basis for CCTV contractors when reporting on their findings.

Most sewers function without problems for years on end but when a blockage occurs it needs to be cleared. The traditional method to clear a sewer block is 'rodding' but now we use 'jetting' as the preferred means. More serious problems, such as severe siltation may need 'winching' to restore the cross section of the pipeline.

We hear many stories today about the problems caused by 'wet wipes' which cause blockages and 'fat bergs' caused by the deposit of oils and fats into the sewers. The clearance of these problems requires special measures and is very expensive. It is generally cheaper to take preventive action than having to dig the deposits out.

ISO 9000

The introduction of ISO 9000 brought new ideas into maintenance and it is now common practice to produce procedures for each item of plant. This is a simple documented set of steps which tell the operator and the maintainer what to do, when to do it and in what order including health and safety issues. An example, which could be produced on a film covered A4 sheet, but is now more likely to involve a QR code, could contain steps such as those listed below:

Example of a Maintenance Schedule

Plant:	Submersible pumping station maintenance

Ref:	MIN/STP/PS/P1and2	Manager:	SA

Created:	24/09/2020	Updated:	2012

Purpose and Scope

Steps to be taken for the weekly maintenance of submersible pump stations

Responsibility

This procedure should be carried out by a trained NVQ3 qualified fitter

Equipment

Pump gang van equipped with tools and clean water jet

Procedure (plant)

- Ascertain who else is present on site and what other procedures are likely to be carried out
- Switch off power feed to pumps and lock out; place 'locked-out' notice in place
- Open the wet well and pull up pump 1
- Examine inlet for signs of blockage and clean with water jet
- Return pump 1 to well and repeat procedure for pump 2
- Switch power back on and run both pumps while breaking up any debris in well with water jet
- Return pump operation to normal duty and standby
- Close well and lock
- Complete station log and gang log with timings

This procedure should be read in conjunction with the 'Submersible Pumping Station Safety Guide' and the Operators' procedure for this type of plant.

Support Functions

Customer Services

The most obvious support service is dealing directly with customers' complaints and queries. It is arguable that, since over half of the issues raised relate to water bills, that it should reside with the Finance Directorate. Obviously customers (we don't call them consumers anymore) want a reliable supply but when it goes wrong, they want the issue sorted and if it can't be done expeditiously, then they want to know when it will be sorted.

In the old days (I do go on about them a bit don't I?) complaints went straight to operators and many were not noted for their bedside manner. Likewise complaints about bills were dealt with in finance and not always in a good manner. Nowadays, virtually all businesses have a dedicated call centre with a team of call-handlers whose job it is to sort out problems. Their remit is, not only to sort out the problem, but to make the customer feel happy when it is done.

Of late, the main problem with call centres has been the long delays with answering incoming calls. "We are currently receiving a high volume of calls but your call is very important to us." This does not ring true (excuse the pun) when it's received from a robotic voice - no matter what time of day you ring up – it simply indicates that the call centre is not properly manned and it gets the customers' backs up.

Public Relations

It's not uncommon for the Chief Exec or MD to keep this under his own direct control as it's often the main thing that the public and customers see when the company is in the news. A professional approach is essential when dealing with media professionals as they need to trust the outlet if they are to treat your output with respect.

The Water Authorities attracted a lot of negative publicity in their early days as the increasing water charges were subject to scrutiny for the first time and pollution in rivers and the sea was bad. Privatisation was also subject to much adverse coverage but things have generally improved over the years – not to say that there haven't been some major issues which have, quite rightly, brought some to account.

Putting out positive messages to the public is commonplace and on-line consultations, such as Severn Trent's 'Tap Chat' can do much to improve the industry's image.

Framework Agreements and Outsourcing

This type of contract came into vogue in the 1980s and has been a favourite with many in the years since. Whilst it has many forms, there is a typical format which consists of a 'framework' without too many specifics. The contract contains basic terms which will include:

- The services that are to be provided
- The term of the contract (usually two years) and any possible extension (usually one year)
- The terms of remuneration and allowance for inflation
- The applicable law and resolution of disputes

Water companies have employed them extensively over the years to cover services that are uneconomic when carried out in-house or when the letting of repetitive contracts is over-cumbersome. Whilst office cleaning was the most common in the early days, we now include such issues as the provision of submersible pumps and design services.

The attraction for the client includes the ease of letting and less need to comply with employment law, whilst the contractor/supplier sees great advantage in continuity and a steady income stream which may be extended if performance is deemed satisfactory.

Legal Services

The water industry, like most utilities had its roots in local government and even up until the 1970s they were run by the Town Clerk who <u>always</u> had a legal qualification. Why you needed to be a solicitor to be a good manager is very questionable but, in days-gone-by, everyone accepted it in the same way that the word of your local GP was sacred.

Now legal services are seen as something that can be bought-in as a service and many companies (especially the smaller ones) 'contract out' the provision of advice on legal matters along with the drafting of legal papers such as contracts.

Transport and Plant

Since the 1974 reorganisation, there has been a gradual movement away from the direct purchase of vehicles towards leasing them from companies who buy them. This is said to have some advantage in accounting terms but the companies generally see it as off-loading a somewhat tiresome (excuse the pun) business onto someone else who is more proficient at it.

Whilst the provision of cars and light vans is usually covered, heavy mobile plant is another issue. Companies tend to run mixed fleets of vehicles such as water and sludge tankers, but hire in plant like JCBs and cranes when needed. Compressors and mobile pumps may be owned or hired as required. Specialist services such as CCTV and sewer jetting are virtually always contracted out.

Emergency Planning

Planning for emergencies was always seen as being part of the operations function as they already handle the day-to-day issues with the networks and plants. However, as manpower numbers have fallen, companies may find themselves unable to respond with adequate resources. Prior to its disbandment, this function was fulfilled by Civil Defence but it's long since gone. Framework agreements are a way to address any shortfall though the contracting company needs to consider what training and coordination needs to be undertaken between the parties to the agreement.

There is one thing (other than death and taxes) that is certain which is that, when confronted with an emergency, 'management' will always tell the media that "this an unprecedented situation." Well it's not. The specific circumstances may, indeed, be unprecedented but the occurrence of an emergency situation is not. One major water undertaking has seen 'major emergencies' in its area on average once a decade:

- The 1947 flood
- The 1953 floods
- Ira bombing of the Elan Aqueduct near Hagley in the 1960s
- The 1975/76 drought
- 1982 bomb threats by Welsh protesters over Elan Valley water
- The failure of Carsington Dam in 1984
- The 'Worcester (or Wem) Incident' of 1994
- 2007 River Severn floods which knocked out The Mythe WTP

Add those experienced in your own area (e.g. Camelford, 1988) and you get the feeling that it's only a matter of time before another one occurs.

The Government has issued guidance on the management of emergencies in DEFRA's 2006: *Planning for Major Water Incidents in England and Wales*. [Why they don't apply in Scotland and Northern Ireland escapes me] It is based on the *Security and Emergency Measures Directive* (SEMD) of the *1991 Water Industry Act*, which requires that water companies provide a minimum of 10 l/h/d for those cut off from the mains.

Whilst we try to engineer-out any possibility of failure, *Murphy's Law* tells us that "*what can go wrong will go wrong*". And it's only a matter of time. Most critical plant is backed up in some way which includes dual power feeds to critical pump stations but not everything can be covered which is why we need planning. Emergency planning involves two main aspects. The first is having the necessary resources available to deal with things when they go wrong. The second is to make sure that those involved in dealing with an emergency situation have the necessary skills. This latter aspect needs to be tested regularly with training exercises to assess the capability of managers and to select the right support staff. Reacting to an emergency situation is not the same as day-to-day operations.

Sewer and Water Main Records

The 1936 PHA required local authorities to keep sewer records though there was no detail in the requirements and, whilst most of the urban areas complied – with mixed results - the duty was widely ignored by others. In the midst of the 1960s slum clearance programme, the City of Manchester had not even commenced compilation of their records which were required to serve the proposed new developments. All water departments and companies had good records of the water mains though not all of them to a common standard.

During the 1970s there was a move to improve sewer records to the same high standards that applied to water mains and a working party was set up by the Standing Technical Committee of the National Water Council. Water mains already had the basis of an accepted specification, set up by an earlier group but sewer records were completely disorganised. The working party produced *STC 25* which set a national standard and also dealt with the troublesome problem with OS maps.

Utilities require base maps to record the location of their pipelines and cables and generally use the OS 1/1250 (twelve-fifty) and 1/2500 (twenty-five-hundred) scale maps. Unfortunately, at that time, the OS was short of funding and had dedicated a twenty-year plus timeframe to producing digital maps. Hence the working party had to come up with a solution which would enable them to deal with a mix of paper and digital records. They called it 'the overlay system' and it allowed for all possible combinations of the base map and the overlaid pipeline record.

Whilst different colours and line-styles are used for to differentiate the functions, details of the pipelines are written parallel to the line itself. The data along a water pipe is: size, material, class and date of laying. For gravity sewers it's: size, material, gradient and date of construction. Manholes and mains junctions all have their own unique identifiers which are related to the OS base maps.

NJUG, which co-ordinated the activities of the utilities, came up with its own simplified specification for the base maps and forged ahead of the OS who eventually responded. They also set up a trial system in Dudley for the digitisation of the records and exchange of them via telephone lines even before we got going with the internet. Nowadays, when you can get the entire OS coverage of the country on your phone, these problems seem to be just part of history.

Searches

Even after privatisation, property searches were carried out by staff at local authorities as they still had agency functions. With their demise, it became necessary for the big water companies to take over this function in the late 1990s and the water-related, property search questions, which used to be on form CON29 were then absorbed by them.

WaterAid

WaterAid was founded by David Kinnersley in 1981 in response to the launch of the United Nations' Water Decade and has brought clean water and hygienic toilets to many since then. Here is a short extract from its website:

Extreme poverty cannot end until clean water, toilets and hygiene are a normal part of daily life for everyone, everywhere. Yet 785 million people live without clean water close to home, and 2 billion don't have a decent toilet of their own. And climate change is making the situation worse.

Women in particular waste precious time walking long distances to collect dirty water. Girls drop out of school because there are no private toilets to manage their periods. And approximately 289,000 children per year do not live to see their fifth birthday because of diarrhoeal diseases caused by dirty water and poor sanitation. We may be called WaterAid, but our focus is resolutely on all three essentials: clean water, improved sanitation and proper hygiene.

There are seven countries in the WaterAid Federation: Canada, United States, United Kingdom, Sweden, India, Japan and Australia. Each mobilises support in its home country, engaging individuals, companies and institutions to raise funds and influence policies to support the charity's mission. All of the UK water undertakers support WaterAid and private enterprises world-wide make significant contributions.

The decade following the formation of WaterAid was 1981 – 1990 and is now known as The First Water Decade which was designated by the United Nations to bring attention and support for clean water and

sanitation worldwide. Since 1981, each decade has seen a re-emphasis of the importance of water supply and sanitation and in 2015 all United Nations member states adopted the Sustainable Development Goals (SDGs), to be achieved by 2030. The goals provide a blueprint to achieve peace and prosperity, now and into the future. WaterAid's aims within the overall goals is to ensure clean water, decent toilets and good hygiene for everyone, everywhere which is embodied in SDG 6 (which WaterAid was instrumental in securing). Clean water and effective sanitation is seen as a vital contributor to other goals such as ending poverty, improving nutrition, securing good health and quality education.

Water break by Scorius (note this is an India Mk1 pump)

Community Water Pump, Tigray by Rod Waddington (note this is an India Mk2 pump)

In practical terms WaterAid provides and installs pipelines, pumps, storage and treatment in addition to offering training in water and wastewater operations and management. Hygiene features strongly in all projects as does sustainability and maintenance of equipment, plant and vehicles.

You can contact WaterAid on-line at https://www.wateraid.org/why-we-are-here and make a donation.

Asset Management

AM has been around for longer than its name as many businesses had systems in place but its recognition as a subject in its own right only became apparent after water privatisation when asset management systems became an essential part of regulation. Its roots in the UK can be found in the grading of sewers produced by WRc in the 1970s but it was mainly developed in Australia and New Zealand in the following years.

The first essential is the existence of a computerised inventory consisting of all of the company's physical assets which is loosely based on the figure below:

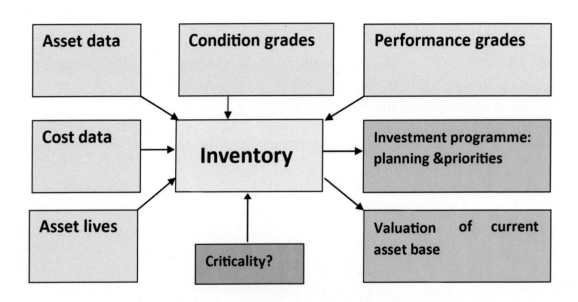

Typical structure of an AMS

After the basic asset information is entered, assets are graded for condition and performance. Replacement cost data and asset lives are added and some choose to include a criticality indicator.

Outputs from the system can form the basis of area and functional strategies which become part of the capital investment programme.

Whilst there is a formal organisation (the IAM) which purports to control the subject, you don't actually need membership to undertake it. An outline of a basic AMS for a water company is contained in *Principles of Asset Management* and a very full description in *SAMS, Simplified Asset Management Systems,* both of which can be found on Amazon.

Design and Construction

Those entering the Water Industry on the technical side tend to be engineers and scientists. Whilst the chemists are most concerned with the processes and water quality, the engineers are the ones who design most of the plant and pipelines. Pipelines are mostly concerned with flows whilst structures and treatment plants require a clear understanding of processes which brings us back to the chemists.

If you aspire to be a competent designer, in the water industry, then you may be required to demonstrate your expertise and understanding in a number of areas:

- Dams and reservoirs
- Potable water treatment
- Water supply
- Distribution
- Sewerage
- Pumping
- Sewage treatment

And, in undertaking these functions you may require some knowledge of:

- Finance and project appraisal
- Asset management
- Operations and maintenance
- Strategic planning
- Town planning
- Building and plumbing regulations
- Ofwat and regulation

Whilst not wishing to present you with a précis of a university course, we hope you will find the following pointers useful.

Design and Construction Process

The Planning and Project Appraisal process which will result in an asset coming into use for the benefit of a company can be summarised in the first three of the following actions:

- Identification of need which may be something of an individual basis or be part of a strategy
- Consideration of options to solve the problem
- Deciding on the preferred option to take forward with identified investment requirements
- Design
- Tender and contract letting
- Construction
- Commissioning

There is a more detail on this subject in the chapter *Investment Planning and Project Appraisal* in: *SAMS, Simplified Asset Management Systems.*

Moving on to the design process, this usually starts with a design brief which tells the designer what the client wants and should specify:

- The objective that the plant or pipelines are intended to achieve
- Their location or the area they are to serve
- The capacity/output
- The timeframe
- Any further details that are relevant to the project for instance, responsibilities

In the case of a plant, there are three main design considerations:

- The process design which sizes each individual process as a basis
- The physical sizing of the tanks, pumps and pipelines and their interconnection
- The detailed structural designs

This latter process is the one which most universities concentrate on at the expense of the first two. It's all very well knowing the formulae for everything but the process and the configuration are just as important. Looking at a sewage treatment plant, each of the stages of treatment needs to be sized and then detailed which will involve the choice of mechanical equipment to be installed. Only after the M&E is sorted, can the structures be detailed.

It is normal practice to carry out a ground survey especially for gravity pipelines.

Once the detailed drawings are completed, the tender documents are drawn up which can be varied according to the type of contract. Some prefer to detail every cubic metre of concrete and ton of soil removed whilst other forms of contract involve less specification and leave the detail up to the contractor. The contract will contain:

- General condition of contract, usually based on one published by the ICE
- The contract drawings
- The specification
- A bill of quantities or a schedule of prices

The contract may be subject to open advertisement or restricted to a select list of contractors who have passed scrutiny and have a track record. After submission, bids are assessed for completeness and correctness and then the successful bidder is chosen based on the method applicable to the contract. Most choose the lowest bid, which tends to lead to a lot of claims but some prefer to take the one nearest the average price. Once selected, the preferred bidder will be appointed and a formal contract signed.

Construction is then undertaken and the contractor is usually paid monthly based on 'certificates' which detail the work completed to date. An alternative is the 'turnkey' contract where the contractor does the detailed design and does not get paid until the project is complete.

There are two stages before final completion. The contractor must work with the client to commission the plant or pipelines and demonstrate that everything works as planned. The contract then enters a maintenance phase which usually lasts a year. If everything is OK at the end of the maintenance period, then the assets are finally handed over and the contractor is released from further involvement. The assets then enter their operational phase and are recorded in the maintenance and asset management registers/pipeline records for looking after by the company.

Case Study – Design of a Submersible Sewage Pump Station

Whilst most would consider this a mundane task or even off-load it on to the pump manufacturer, the actual process is indicative of the need for a logical procedure in any design process which should avoid actions having to be repeated. The mnemonic was the winner of a competition on a BTEC course in 1990 and the prize was the cricket ball which was passed through the pump impellor which had been displayed.

1. Local   Determine the **location** of the station which will be at the low point of the site
2. Flygt   Calculate the **flow** which will come from the connected properties
3. Man   Size the force **main**, limiting the velocity to 1m/s in the main
4. Has   Calculate the **head** by adding the static and friction heads together
5. Pump   Get the size of the duty **pump** from the manufacturer's catalogue
6. With   Design the wet **well** for the required number of pump starts per hour
7. Seals   Detail the **site** and the valve chamber plus access for maintenance

A check on the retention time in the force main is useful if septicity is likely to be a problem.

Prior to this procedure, most designers just sorted the layout and the flow but left the pumps to the mechanical engineers. It frequently led to the installation of grossly over-sized pumps and consequent septicity problems. There has been much debate over whether such stations are better with three (smaller) pumps rather than two larger ones. The normal choice is to settle on two.

Materials

A company that handles any fluid, such as water, will be heavily dependent upon its pipelines so we start there and deal with the other stuff later.

Pipes

The first thing is to understand about pipes is the preferred means of manufacture. They are either spun or extruded. Spun pipes are made using centrifugal forces to create the pipe within an external mould. Extruded pipes, especially plastics, are forced, under pressure out of a special nozzle and cooled as they are extruded. This means that they may be much longer than spun pipes which are limited by the size of the mould.

Asbestos Cement (AC)

Originally developed in Italy, and first used in the UK in 1921, AC was still being used into the 1980s but, following concerns about the dangers of asbestos dust, their use was discontinued. If handled correctly there is no danger to personnel and there is no evidence that consumers suffer any health issues due to their use. In addition to dubious concerns about health hazards, the introduction of newer materials, specifically plastic and longer-life iron pipes, AC has been phased out though there are no current programmes to replace existing water mains simply because they contain asbestos.

Cast or 'Grey' Iron Pipes

Cast iron (CI) pipes date back to the industrial revolution when they were used to replace wooden pipes. In order to ensure adequate strength and to cater for casting imperfections, they have relatively thick walls, sometimes up to 50 mm for the larger diameters. This thickness limits the long-term effects of corrosion of the iron and the overall integrity of the pipe. There are many cast iron pipes, especially larger diameter trunk mains, still in satisfactory use today.

Spun iron, as opposed to cast iron pipes were first produced in the UK in the 1930s. This entailed better uniformity of the wall thickness and, thanks to the spinning process, denser iron. However, grey iron has a major flaw - the molecular structure includes graphite flakes which produce a brittle, more rigid material that is easily damaged if dropped or hit during laying operations.

Ductile Iron (DI)

An improvement was made to grey iron by adding magnesium to the molten metal, which was then spun in the same way. This greatly increases the strength of the material which changes from a brittle, easily damaged, metal into a tough, more flexible material. Being stronger than grey iron, the pipe walls are thinner. This produces lighter-weight, easier-to-handle pipes but corrosion can quickly affect the thinner pipe wall resulting in leakage and reduced service. To counteract this, ductile pipes are protected internally and externally, traditionally with bitumen or, later, with synthetic materials. Although not used as much as it used to be, DI remains the material of choice for high pressure pumping mains and where polluted ground conditions excludes the use of plastic pipes.

Steel

Steel pipes and fittings are popular all over the world, especially where there is an oil industry. However their susceptibility to pin-holing has made them unpopular in the UK except for very high-pressure applications where they need both external and internal protection. Up to about 500 mm diameter, the pipes are made from ingots of hot steel to form a seamless pipe. Unlike other materials, larger diameter steel pipes are made by bending steel plates to form a circle and butt-welding along the seam. Because of its strength, steel pipes can have significantly thinner wall thicknesses than iron or plastics for any given internal pressure. However, if the pipe wall is too thin then the pipe can become distorted and take on an oval shape. Steel pipes can be joined by butt-welding, welding sleeve joints, Viking Johnson couplings or welded flange joints.

Viking Johnson pipe coupling

The main problem with steel, as with iron, is corrosion, both external, due to ground conditions, and internal from the transmitted water. Being much thinner than iron, corrosion can render the pipeline unserviceable, sometimes within a few years of laying in aggressive ground. Both internal and external protection is essential in order to ensure an economical life but protection raises the overall cost considerably. External protection consists of either a priming coat covered with bitumen sheathing, bitumen enamel or coal tar enamel wrapping. Internal protection comprises either cement-mortar, epoxy resin, polyurethane coating or bitumen with a suitable primer to ensure adhesion.

Some consider that the best internal protection for both steel and iron pipes is cement-mortar. This is applied in the factory by spinning the pipes at high speed and introducing the mortar as a liquid slurry. A variation of this method is used to rehabilitate existing pipelines in-situ.

Pre-stressed Concrete

In the UK, pre-stressed concrete pipes are rarely used and then only for specific high-pressure applications.

Concrete and Reinforced Concrete

Unreinforced concrete pipes are rarely used in the UK. RC pipes use a mild steel cage for reinforcement and are spun in a rotating metal cylinder. The spigot and socket joints are part of the mould and the pipes, which are used more for sewers than for water mains, are jointed using a rubber ring. They are more suited to gravity pipelines than pressurised ones.

Clay Pipes

Clay pipes have served as the backbone of the sewer industry for most of the small diameter gravity pipelines though plastic pipes have become more popular of late due largely to their need for less jointing. Vitrified clay pipes had a glaze which made the clay impervious to water but more modern methods of manufacture mean that the baked clay is resistant to decay and glazing is unnecessary.

Brick Sewers

In Victorian times, manpower was cheap and so was brick so many sewers were laid, sometimes in shallow tunnels using bricks to line the tunnel. Whilst some were circular in cross section, many were constructed as 'egg-shaped' sewers as this was perceived to provide better hydraulic characteristics. Many survive today, depending on the quality of construction and especially on the type of mortar used to bind the bricks together. Lime based mortars soften when in contact with water and many sewer collapses have been caused by the loss of just a few bricks. Some of these sewers have been refurbished due to a technique which places a plastic or glass fibre lining inside the line. Many main and interceptor sewers were constructed in brick but, generally to higher standards with multiple rings which have enabled them to continue in use – and hopefully for many years to come.

There are also many brick culverts, especially where minor waters courses cross under roads.

Glass Reinforced Plastic

GRP is used in sewers or water mains in special applications such as short tunnels and culverts.

Lead

In the UK we tend to think only of service pipes being made from lead but it's thought that the Romans were the first to use it as a pipe material and they produced pipes ranging in size from about half an inch to 22 inches. It has always been suspected that using lead as a conduit for water is hazardous to health and so, by the twentieth century, lead was phased out for new installations and replacement programmes were initiated. Lead is more dangerous to health in soft-water areas so various strategies have been undertaken to mitigate the potential hazards. The common solution is to replace all of the lead pipes and this is included in a programme of overall improvement which often includes the installation of meters. Where there is a preponderance of lead pipes in an area, then short-term mitigation measures can include pH correction using lime dosing at the treatment plant.

Copper

Copper pipes are not generally used for supply but are common everywhere in internal plumbing.

MDPE and HDPE

New supply pipes for domestic use in the UK are now made from MDPE and are generally 25 mm diameter. The stated 'size' of PE pipes is now the external diameter but previously pipes were described by the internal diameter, or bore of the pipe. As an example, the bore of a 25mm diameter MDPE pipe is actually 20.4mm. Whilst MDPE sets the standard in the UK, HDPE, which is slightly less flexible and hence more prone to damage during laying, is only used where higher pressures are needed. It is commonly used abroad.

Pipe Joints

If a pipe line is going to fail, then it's ninety percent certain that it will be at a joint. It's quite rare, though not impossible, for the joint to be stronger that the body of the pipe. They come in two main kinds – 'push-fit' and 'mechanical' though the use of plastics has brought about solvent welding, butt fusion and electrofusion as means of jointing.

Originally, metal water pipes were joined using 'run lead joints' whereby liquid lead was poured into the joint in a single pour and prevented from flowing away by a clip around the joint. The lead was prevented from running into the pipe by packing yarn into it prior to pouring. Once cooled and solid, the lead was caulked using a hammer and chisel to ensure it was packed tight. These joints can leak as a result of ground movement, but for well-laid, carefully jointed pipes they have proved to be extremely reliable and long-lasting. Many water mains, especially large diameter trunk mains, are still operational with run-lead joints. Socket and spigot jointing is still the preferred method for metal pipes because it's quicker and easier than using bolted couplings or separate collars.

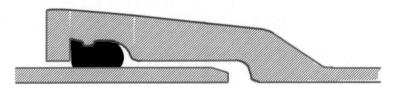

A Tyton joint as used with pipes previously manufactured by Stanton and Staveley in the UK and other pipe manufacturers world-wide.

To prevent push-fit joints coming apart in pressurised systems, concrete thrust blocks are constructed at bends and tees in order to prevent movement. In-line sluice valves are joined to pipes usually via flanged adaptors and the valve is restrained within a concrete or brick wall embedded each side of the trench. Iron pipes have their own design called the 'Tyton' joint which has also been adapted for other materials. Most concrete and clay gravity sewers use a 'spigot and socket' joint which has a rubber ring to provide the seal.

Brick

Prior to the widespread use of concrete, brick was generally the preferred material for construction and remains so for housing. Its use for culverts and larger pipelines necessitates some internal formwork which would normally be a temporary staging made of timber which is removed after the mortar has set. Where brickwork is in contact with water, or even dampness, it is essential that 'engineering brick' is used to avoid deterioration. These bricks are made from special clay and baked at a higher temperature to ensure a waterproof skin on the outside. Staffordshire Blue engineering brick is famous worldwide.

The early mortars used lime as an agent and are susceptible to the mortar leaching out in contact with water. This has been a problem with many old brick egg-shaped and circular sewers which have required replacement or lining. Many of the Victorian interceptor sewers were constructed with multiple rings of brick and remain in use today. Bazalgette's embankment interceptors in London are a prime example.

Concrete

First let's get some basics sorted – how are cement, mortar and concrete related? Portland cement is a fine ground powder which has been baked to remove all of its water. Its manufacture is the third largest atmospheric carbon producer of any industry due to the power consumption in manufacture. When water is added, it undergoes a chemical reaction (see Wikipedia for details) and hardens though it has little use unless mixed with other ingredients. So when we say that we construct something using "cement" we are only half correct.

2006/04/21

If we add a soft sand ('builders' sand') to cement in a prescribed ratio (say 4 to 1) we get mortar which we use to lay bricks. The bricks provide the strength and the mortar joins them together. Lime mortar, which used lime instead of cement, is rarely used today.

If we add a sharp sand and stone aggregate, again in prescribed proportions, we get concrete. The ratio 4:2:1 - stone aggregate: sharp sand to cement is typical. The making of concrete is a science in itself and, again, Wikipedia can enlighten you if you want to delve further.

Concrete Housing Construction by Concrete Forms

Reinforced concrete is when we add mild steel reinforcing bars, in the desired shape, to the structure in order to give it tensile strength in addition to the compressive strength provided by the concrete. Whilst it was normal to mix concrete on site for any major project, it is now more common for it to be delivered to site by special tankers which keep the mix in motion during travel. You only have to specify what you need and the supplier looks after the design of the mix.

The use of RC is quite complex as the concrete starts off in a liquid form before solidifying around the steel frame which provides the tensile strength. In order to stop it flowing, you need 'shuttering' which is a timber mould forming the exterior shape of the desired structure. The hydraulics involved in a set of treatment processes can be quite complex and the shuttering has to reflect this. Many concrete products are pre-cast such as manhole rings.

Neatly stacked pipes by Carlfbagge (actually they look more like manhole rings)

Steel

There are two main kinds of steel that are used to build treatment plants and facilities such as pump stations. We have referred to mild steel reinforcement for concrete structures above and this makes up a large proportion of steel usage when constructing a treatment plant. Every junior engineer has to learn how to compile a 'bending schedule' for the contractor to be able to make up the bent shapes of the steel rods for assembly prior to pouring the concrete.

Most other steel is in the form of products which are purchased from suppliers or manufacturers. Valves and other appurtenances tend to have iron bodies with nuts, bolts, stems, etc being stainless steel. On a treatment plant, much of the machinery which is essential to the process is made from steel and acquired from specialist makers. They will be responsible for its design as well as manufacture but the client needs to specify or approve what protection against corrosion is required. Cathodic protection is a specialist subject in its own right.

Most access points and manholes require covers which are made of iron or steel which is protected against corrosion.

Stone and Masonry

I probably should have put this first but, due to its lack of use in more modern times, I actually forgot about it. What's the difference between 'stone' and 'masonry' – stone is the raw material taken from the ground and masonry is it when it's been worked on by a 'mason' into a finished product and installed as part of a structure.

Craig Goch by Mike Pennington

The Greeks were very keen on their structures and took ages to build them as they were nearly all made of solid stone. The Romans were much speedier as they often constructed the outer faces of dressed stone (masonry) and then filled the interior with concrete.

The Victorians used both brick and masonry extensively and their dams are a prime example as many were constructed entirely from stone. When Birmingham City Water Works planned to build an even bigger dam in the Elan Valley in the late 1940s, they decided, in order to blend in with the other dams, to face it with masonry. It's said, that when Claerwen was opened by Princess Elizabeth in 1952, that the stone facing doubled the cost of what is actually a concrete dam. The interesting thing is that they actually built it the opposite way round from the Romans as they laid the concrete first and then added the facing. Being the largest of the Elan dams, it's well worth a visit, especially when it's overflowing.

Princess Elizabeth at the opening of Claerwen Dam in 1952 by Geoff Charles

If you remember a Top Gear episode with Richard Hammond winching a land Rover up the face of a dam – that's Claerwen.

Dams and Reservoirs

Dams and reservoirs are closely related as the 'dam' holds the water back and the 'reservoir' is the valley/area above it which contains the water which is 'reserved' as a resource. Dams are normally sited along the course of a river or stream and are constructed to block off the valley and only allow a specified amount of water to pass. They come in two principle kinds – earth dams and concrete/masonry dams.

A flight to the Derwent Dams by John McLinden

Reservoir Water

Large raw-water reservoirs store water to ensure continuity of supply during seasonal dry-spells. However, they also help in the removal of harmful organisms. Companies ensure that, as far as possible, the water entering a reservoir is as pollution-free as possible but accidents can, and do happen. Bacterial pollution can enter a water course from farming or industrial accidents, sewage treatment plant failures, storm sewage overflows or accidents involving chemical-carrying road tankers. Even after such incidents there is a significant decrease in the levels of pathogens after a few weeks of storage. This 'treatment' relies, primarily, on the effects of ultraviolet light along with the sedimentation of pollutants.

Water quality in a reservoir varies with depth although there are no hard and fast rules as to which is best. Water temperature plays a role, especially in deep reservoirs. The water is warmest nearer the surface (epilimnion) and colder further down (hypolimnion) with what is known as a thermocline or temperature gradient between the two. Seasonal weather variations, or even sudden changes of air temperature, can cause significant changes in water temperature and if the epilimnion becomes cooler than the hypolimnion the water will 'turn over', causing sedimentation to rise up and reduce the quality of water near the surface. To ensure that the best quality water available is transmitted to the treatment plant, staff constantly monitor the quality at varies depths and the infrastructure is designed such that water can be drawn from different levels.

The Components of a Dam

It's easy to think that a dam is a simple structure placed across a watercourse to hold the water back and this is largely true for most small structures. However, there are a number of things to consider which include:

- how much water is to be held back?
- How does the watercourse continue to flow when the dam is full?
- What will happen if there is a flood and the dam overtops?

Dam designers take these issues into account and have build in a number of features so now a modern dam will have these things as standard:

- A water-tight core
- Upstream and downstream faces or embankments to ensure the stability of the core
- An overflow for normal operation
- An emergency overflow for extreme conditions to prevent over topping
- A scour valve to enable the reservoir to be drawn down
- A by-pass pipeline and valve system to release water to the downstream watercourse and to provide 'compensation water,' i.e. water which is required to maintain a specified flow in the river to mitigate the potential loss of riparian rights
- A draw-off tower
- Monitoring equipment to ensure that nothing untoward is occurring within the structure

This last item is particularly important for earth dams as any movement of water though the embankments can result in failure.

Draw-Off Tower

If you visit a reservoir you'll most likely see a structure sticking out of the water, it's often near the dam but can actually be anywhere although its best location is where the water is deepest. Not all reservoirs have a draw-off tower, especially relatively shallow ones. In such cases it goes through the dam itself and therefore there's no separate draw-off arrangement.

Depending on the depth of water stored at any one time and the quality of water at varying levels, there is a need for draw-off towers to incorporate several outlet pipes, each at a different level. There may be as many as seven such levels or as few as two; the number being proportional to the maximum depth of water. The draw-off system generally comprises a vertical pipe within the tower itself. In traditional, sluice-valve controlled systems there will be remotely controlled (usually electrically operated) valves spaced along the length of the standpipe.

Some draw-off towers use siphons instead of valves and these are built into the tower wall, facing the reservoir water. The advantages of siphons compared with valves include:

- Simpler construction
- No moving parts
- No leakage through faulty valve seatings
- No requirement for access to the valves
- Less maintenance

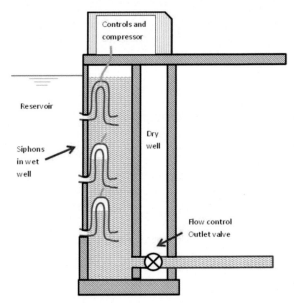

Concrete and masonry dams

Whilst concrete has been around for centuries, it was rarely used for dam construction until recent times. The Victorians preferred to use masonry and constructed very elegant dams, many of which can still be seen today. Stone was normally sourced locally and a large number of masons employed in quarrying the stone, shaping it and placing it in the required position to make the structure which is naturally water-tight.

Earth dams

Earth dams are quite different in that they are designed to blend in with the local topography and also, that they are not naturally water-tight. In order to achieve this they have a clay core which is constructed at the same time as the embankments which are there to support it on the upstream and downstream faces. All kinds of dam require extensive and detailed ground investigations to ensure that they have sound foundations. Any shortcomings in this regard can result in failure, as occurred at Carsington in the early 1980s.

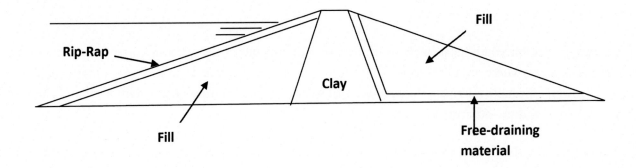

Some reservoirs have only a small catchment which means that it could take years to fill and the resource would be small. Thus we now have 'pumped storage' schemes. In this set-up, water is pumped from a lower source into the reservoir for storage. Carsington Reservoir has quite a small catchment in the Scow Brook so gets most of its water from the River Derwent at Ambergate during periods of high flow. It then releases water back to the river during periods of low flow or via a tunnel to Ogston Water Treatment Plant for domestic supply.

Foremark Dam under construction

Hydropower

In Scotland, where the topography is more challenging, much of their energy is generated using water as a power source. Whilst it may be simply a matter of having generators powered by the outflow of water from a dam, some use the water for generation to cover peak demands and then pump it back when there is excess power in the grid.

Case Study – Cruachan Pumped Storage Power Station

Built in the early 1960s, Cruachan actually uses more power than it generates but its purpose is to support the grid at times of high demand and take power back when the grid is in surplus. Due to its design, it is able to respond within seconds to a surge on the grid, for instance, when Coronation Street finishes on TV and everyone puts the kettle on at the same time.

Kilchurn Castle Loch Awe by ronniefleming

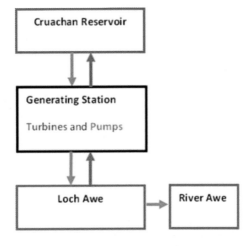

Design of Treatment Plants

The design of a treatment plant has surprisingly more steps than you would think and starts off with a few basic questions:

1. What's the purpose of the plant? – e.g. to provide potable water to the town of Everywhere (population 50,000)
2. Update the project estimates (from the wild guesses made earlier?)
3. Where should it be located? – e.g. on the hill overlooking the town
4. Where will the water come from and how will it get there?
5. What processes shall we use?
6. Draft plant schematic
7. Design intake
8. Design process A; design process B; design process C
9. Design outlet and storage
10. Size interconnecting channels and pipes
11. Pass drawings to M&E to size pumps
12. Pass drawings to plant suppliers for their input
13. Do detailed structural designs and drawings
14. Prepare bills of quantity
15. Prepare contract documents
16. Advertise and compile select list of tenderers
17. Select main contractor
18. Supervise construction
19. Supervise commissioning
20. Confirm final costs
21. Operate and maintain,
22. Satisfy customers

Design of Pipelines

We design a pipeline based on two principles:

- How big does it need to be to carry the desired flow and

- How strong does it need to be to withstand the loads on it?

Water Mains

Water mains are always pressurised so they can follow the ground, usually at a constant depth of around 600mm below the surface. Generally speaking, there is little need to consider the external load on a buried water main as the internal pressure will be the crucial factor and they are generally at a shallow depth compared with sewers. So, the policy with respect to maintaining pressure in the system will be what matters and this can be a political decision rather than an engineering one.

In Israel, the distribution system is maintained at a very high head due to its (perceived) need for fire fighting. In Russia, the enormous tower blocks, which are a feature of the communist system of government, are fed by a roof-top tank which is fed directly from the mains. Obviously such systems require very high pressures in the distribution system. Most countries rely on any extra pressure for fire fighting to be provided by the fire tender as it extracts water for the main and tower blocks are served from a break tank at ground level.

In a fairly flat area, all of the mains may be of a similar material and designed to withstand a nominal pressure with a large safety factor built in as this does not greatly affect the cost of the pipes. A minimum pressure of 15m head in the mains is normal which equates to 1.5 bar.

When we start to consider the ageing process then the effects of water on unlined metals becomes crucial. Unlined cast iron pipes will rust and decay over time and require heavy maintenance or replacement to keep up with demand. Plastic pipes have no such issues but unprotected steel is a definite no-no for water mains as it fails by 'pin-holing' after time.

With most pipelines, it is the joints which create the most problems and this was a major problem when we still relied on coal. After a two metre seam of coal was abstracted the overlying land would subside by at least a metre after a relatively short time. This subsidence wave, as it progressed across the countryside, would stretch and compress the pipes putting a severe strain on the joints whist leaving the pipe body intact.

The preferred modern materials depend largely on the size of the main, plastic (MDPE or HDPE) for smaller mains and ductile iron for larger ones. Fibreglass is occasionally used for special applications but is not common. Steel is best avoided.

The size of most distribution mains will be simply 100mm diameter; though mains which are able to serve for fire fighting will normally be at least 150mm diameter. The size of supply and trunk mains is determined according to the calculated maximum demand and the need to economise by restricting the friction losses in the pipeline. This means keeping the velocity to a minimum.

Force Mains

Force mains are designed as if they are water mains i.e. pressured pipelines and are usually laid at shallow depths. Whilst a high factor of safety is always used, it may not be sufficient to prevent bursting when the pipeline is subjected to 'surge pressures'. Like 'water hammer' these sudden variations in pressure occur when a pump shuts down and the outlet valve to the force main slams shut. This results in a sudden reverberation of pressure within the main which subsides within seconds. We can use a slow-closing valve, which reduces the pressure gradually rather than abruptly, to avoid this phenomenon.

Force mains are normally sized to keep the velocity of the pumped sewage in the main below one metre per second in order to minimise power consumption.

The retention time within a sewage force main is crucial in considering whether there will be a septicity problem in the main. This results when all of the available oxygen is taken up and the sewage becomes 'septic'. When it reaches the outfall of the main, the resulting release of sulphides can lead to a build up of sulphuric acid which can attack pipes and structures especially if they are made of cementatious materials. The bacterial agent responsible for this is aptly called 'bacillus concretivorus' which does what it says on the tin.

Gravity Sewers

Unlike pressurised water mains, gravity sewers have to be laid at a constant gradient and in a straight line between manholes. They are usually laid at a depth of around 2.0 metres above the invert of the pipe. This is to enable drains to be connected by gravity but it also provides some level of protection to adjacent properties in the event of surcharge resulting from a blockage or flooding.

Before H&S became mandatory, deaths were common amongst construction workers who laid pipelines. Severe penalties are now made against companies and/or individuals who are held responsible for the injury or death of anyone on site. No specific depth-related rules are now in place regarding the requirements for trench timbering but it is always wise to err on the side of caution by battering back trench sides and/or provide timbering.

The sizing of gravity sewers depends, very much on their function. Foul sewers are designed to carry at least 6xDWF and are normally at least 225mm diameter which is to minimise the risk of blockages. Some relaxation to 150mm has been permitted of late. It's just a matter of counting houses and multiplying by the average water usage (135 l/h/d) and the average occupancy (say 2.5).

Storm and combined sewers are more complicated as we need to make assumptions about the rainfall that will enter the pipeline. The contributing area is drawn up on a map and then the impervious area is shaded. This is converted to a fraction to be multiplied by the total area giving the net area which will contribute rainfall.

The 'rational method' (aka Lloyd-Davies) then needs the total time of flow that the storm water will take to reach the end of the pipe which is being designed. A look-up table of rainfall intensities is then used and the two factors are multiplied together to give the runoff. Having determined the gradient of the pipeline, the Colebrook-White tables enable the required size of pipe to be determined. The whole process (except for the Tables) is very arbitrary but it works, having taken no account of the fact that, if the capacity of the pipe is exceeded then surcharge of the system will provide a built-in safety factor.

Whilst manual methods are still adequate for most systems, modern computerised methods are now more common. Early versions used standard rainfall profiles which generated the flow in the system, taking account of the in-built storage as the flow builds up. Even more applications now use actual rainfall profiles to mimic the flow in the computerised model.

The structural design of the pipeline is based on a 'narrow trench' as this is how we attempt to construct most sewer pipelines. Generally the aim is to follow the topography and keep the pipe at a depth of around two metres. This is to allow adequate depth for the drains to be connected. Types of bedding are specified as the laying of gravity sewers needs to avoid points of stress (it was common practice in days gone by, to prop up the pipes, at each joint, on house bricks resulting in frequent failures).

'Type B bedding' is commonly used and requires that the pipe be laid on granular material and then the trench be backfilled with more granular material which is carefully consolidated. The constant gradient of the pipeline is maintained using level boards at each end of the length to be laid, and boning rods.

Tunnelling

Water engineers, especially the dirty kind, can be a little like troglodytes as they spend a lot of their time where the sun don't shine. On occasions they are confronted with an apparently unmoveable mountain and have to decide how to get round it or through it. The answer's a tunnel – what's the problem - well that's the way it seems if you are a tunnel engineer. Mining has been around for aeons and miners were the people who first built tunnels – in order to get to the mineral that they were looking for, and usually, following a seam. Water engineers have always used tunnels as it's a long term solution which avoids pumping and the resulting revenue costs.

The tunnelling shield was invented in Victorian times and this enabled some of the great engineers of that time to construct things which had not been thought of before. The first Harecastle tunnel for the Trent and Mersey Canal took eleven years to construct overseen by James Brindley and it was superseded by a parallel one built fifty years later by Thomas Telford.

The northern end of the Harecastle Tunnels, near Kidsgrove, Staffs. by Martyn Wright

After the canals, tunnelling expertise was extensively employed in building the railways as, like the canals, their grading is all important. As water became more important, especially for industry, and new remote sources were found, tunnelling became an important part of aqueduct construction. The Elan Aqueduct crosses some 73 miles of undulating countryside with a variety of constructions and a duplicated tunnel as it passes through Clee Hill.

In the early days, tunnel construction copied the pioneers and they were dug out manually and lined with brick as that was the only available material that was suitable. In the mid 20th century modern methods became established and 'tunnel boring machines' (TBMs) became available which meant that a lining of concrete segments was used and this remains the accepted means of construction today.

When a short tunnel is required, say, for instance, to get under a road or railway, then there are modern methods to bore under the obstacle and line the tunnel with pipes or even plastic linings. This is called 'thrust boring' and is now commonplace.

Before the advent of mechanical diggers, it was often as cheap to construct a shallow tunnel under a city street as to use open trench construction. Many egg-shaped sewers, in cities such as Manchester, were tunnelled and lined with brick as the tunneler proceeded. Many of the better examples are still in use today though many have been lined with glass fibre or plastics.

Land Drainage and Irrigation

Internal Drainage Boards

IDBs are a kind of authority, covering an area where drainage is problematical. Their history dates back to 1252 in the reign of Henry III. They have permissive powers to undertake works of drainage and manage water levels within drainage districts. Their area is not determined by other council boundaries, rather by water catchments. They are located in The Broads, Fenland, Lincolnshire, The Somerset Levels and low-lying Yorkshire.

Yaxley Fen by Julian Dowse

Dykes and Ditches

Those that manage the low lying areas have many small watercourses to manage – the dykes and ditches, many of which are below the level that they drain to and therefore they require to be pumped out or drained at low tide.

Irrigation

When we talk (or write, even) about the 'water industry' we tend to concentrate on the provision of potable water and its collection and treatment after use. But we actually use far more water to irrigate the crops which we grow and the statistics, which show how much it takes to grow, for instance, an apple, are truly startling. Whilst we dread the effects of a prolonged drought on our water based habits, it's the farmers who bear the brunt of a prolonged lack of rain (and do they moan about it!).

K-line Irrigation System by USFWS

Most arable land in the UK is able to sustain crops in during normal conditions without intervention. However, when you see the rain guns shooting water over a field, you know that it's dry. Whilst some farms keep ponds to feed their modern irrigations systems, much of this water is pumped directly from the adjacent watercourse.

Modern intensive systems use even more water and garden centres are major consumers.

MAFF and Food Production

The Ministry of Agriculture, Fisheries and Food (MAFF), was established under the Board of Agriculture Act 1889. It was preceded, by the Board of Agriculture which was founded in 1793 as the Board or Society for the Encouragement of Agriculture and Internal Improvement - which lasted until it was dissolved in June 1822. MAFF was absorbed into Defra in 2002.

The objective of these organisations was to improve food production by providing information and advice plus help with funding for related projects. Their concentration was, therefore, on two opposing issues – land drainage and irrigation.

Agriculture by Mripp

In recent years, MAFF came in for serious criticism as they had virtually canalised many of the nation's rivers without too much regard for the environment and the associated wildlife. A river, under MAFF's governance was likely to be straight and devoid of trees. It's said, that a water authority employee, when giving a talk on the role of MAFF in the early 1980s showed a slide of a lower reach on the River Trent and peered closely at it. He then said, to the dismay of the attendees, "It appears that someone's allowed a tree to grow there."

Fortunately, due to the efforts of environmental lobbyists, these policies are now discredited and discontinued.

When "Coal was King" there were a whole series of power stations along the length of the Trent which all took their cooling water from the river and discharged it back at a higher temperature.

The power stations that used the river as their source of cooling water are/were: Meaford, Rugeley, Drakelow, Willington, Castle Donington, Ratcliffe-on-Soar, Wilford, Staythorpe, High Marnham, Cottam, West Burton and Keadby. It was once called "Megawatt Valley" as they produced a quarter of the country's electricity.

River Trent at Sawley by Samuel Mann

Rivers

The Ages of a River

We all know about rivers don't we? Well yes and no, because when something goes wrong then we all jump in as experts when we actually don't know as much as we think so let's start from the beginning.

Even before a rivers starts, we have what's called a 'watershed' and this is the divide between one 'catchment' and another. If you've ever been to Dudley Castle and Zoo, then you will have seen one of the best examples. Stand on the wall of the old castle and, to the east is the catchment of the River Tame which meets the Trent and hence flows out to the North Sea via the Humber. But looking to the west, you overlook the catchment of the River Stour which joins the Severn and hence the Atlantic Ocean via the Bristol Channel. Whilst simple in principle, it has become problematical as our friends across the pond have confused the words and now use 'watershed' in place of 'catchment'. [Don't fall into this trap yourself]

A river starts with a spring, usually on a hillside or at the bottom of a hill and that starts the water flowing in a brook or a steam. Not sure what the difference is but a brook is somewhat more poetical and they babble whereas streams don't. The stream becomes a young river and, in these upstream stretches we often get waterfalls. As we move downstream into the middle sections of the river, it becomes more sedate as the land is flatter and the river often meanders. As it approaches the sea, our river becomes an estuary and it may be tidal.

Young Upper Reaches

It's a matter of commonsense (or physics, if you prefer) that streams start in the higher ground and flow downstream. The flow thus tends to be turbulent which means that it has the power to carry sediment and even move rocks. Thus the upland sections tend to be more volatile, they flow faster and are more powerful thus the time taken for water to flow is much less that when it gets into the lower reaches. Some of the upland areas are covered in forest and many are in attractive countryside, national parks, even.

Middle Reaches

As we progress downstream, the river widens but is still fast flowing until it becomes more sedate and the flow less turbulent. The river will often flow through its own 'flood plain'. This is a flat area which allows flood water to spread out over a large area which, in turn, reduces the severity of flooding downstream. Whilst this has been understood since the beginning of time, some of us still buy a house in the flood plain and then complain when it floods.

Lower Reaches

As we progress downstream, much of the course tends to be through agricultural areas with more fields and fewer forested areas. As there is more arable land, there tend to be a higher risk of pollution from fertilisers containing phosphates and nitrates - the River Wye is suffering as I write. As the river widens, it flows through its own fluvial deposits which are easily moved during floods. Thus we have meanders and oxbows as regular features.

Navigation

The middle and lower reaches have supported navigation with boats since the middle ages and before. Many have been improved by the construction of weirs and locks which alters the nature of the flood flows. A natural river tends to have turbulent flow but when we construct a weir, we hold it back so the flow becomes laminar and hence suitable for navigation.

Dredging of the channels is normal but its rarely carried out for drainage purposes – it's for navigation - to keep the bed of the river at a depth so that boats can pass without bottoming. The construction of the weirs is a hindrance to flood flows so they need to be carefully designed and some, as on the Thames, have variable gates on them.

Estuaries and meeting the Sea

The river then gets close to the sea and becomes an estuary. Most will be tidal and thus the area may experience twice-daily changes in its status with regard to the water which tends to be brackish. Some of the best wildlife areas in the country are estuarial and support a wide variety of birds. The larger estuaries normally support navigation as they lead to major cities like London and Liverpool.

Tidal Bores

The gradually tapering shape of an estuary means that the rising tide is funnelled by the converging banks resulting in a wave which gradually gets higher. One of the best places to view a bore is Minsterworth in Gloucestershire where the better bores will have surfers on them and some have ridden them for miles. There's a timetable on-line which tells you when the bores are expected and roughly high they will be.

Severn bore surfing by PapaPiper

We all know about the Severn Bore but virtually all rivers have their own bores at regular intervals. The bore on the River Trent was named the Aegir (which means 'sea') by the Danes and has kept its name ever since.

The River Trent Aegir by Lincolnian (Brian)

Trout Farming

Bibury Trout Farm has been open since 1902 and uses pristine Cotswold water to support a thriving business breeding and selling rainbow and brown trout. Other trout farms are available.

Bibury Trout Farm by Karen Roe

What's in a name?

So, the story goes – that when the Romans arrived, and wanting to keep everything in order, they asked the inhabitants what the local river was called and got the answer "abon" which is Celtic for 'river'. This explains, if you extend the logic, why we have so many English rivers with the same name:

Avon (9) in Bristol, Devon, Warwickshire, Hampshire 2, Gloucestershire 2 (both called Little Avon) plus 3 in Scotland

Stour (5) in Dorset, Kent, Suffolk, Warwickshire, Worcestershire

Ouse (4) Little Ouse in Norfolk and Suffolk, Great Ouse in east Anglia, Ouse in E + W Sussex, Ouse in Yorkshire

Derwent (4) Cumbria, Derbyshire, Northumberland and Yorkshire

Flooding

This text is based on an earlier paper - *The Gospel According to Noah* which may be found by Googling it.

Virtually no-one (with the exception of the Environment Agency) bothers about flooding until it actually happens and it is always someone else's fault even if you did buy a house in the floodplain. However, one other thing is clear – that preparation, by both agencies and individuals, can be a great help in mitigating the effects of almost any extreme event. The roles of the various agencies are widely misunderstood so they are listed later on in this chapter.

Planning for an emergency can be conveniently split into three main sections:

- Identification of areas at risk
- Advance planning with hard and soft measures
- Dealing with the emergency
- The aftermath

Flooding comes in very different types which must be identified. These are generally categorised as follows:

Coastal flooding may be onto low lying land which will normally present a problem during high tides. It may also occur onto protected seafronts during heavy storms and tidal surges. Damage is often caused by material picked up in waves and by the combination of wind and wave action. Tsunamis are a very severe

type of coastal flooding, usually caused by earthquakes or volcanic action underwater.

Tidal Flooding occurs when a high tide exceeds the normal level of the flood defences, usually during 'spring tides' combined with low pressure and a prevailing wind that blows up an estuary.

The 1953 floods were a case in point when a severe low pressure settled over the North Sea which raised the sea level, even before the high tide which destroyed sea and river defences causing severe flooding along the East Coast and into the east of the country.

To avoid a repetition, and flooding in London, the Thames Barrier was constructed to prevent tidal surges from entering the upper part of the Thames Estuary. In addition, the defences along both sides of the estuary were consolidated to a level commensurate with that of the barrier.

Fluvial flooding is largely concerned with major rivers in the flood plain. It is usually the result of prolonged rainfall, melting snowfall or a combination of them (as in England in 1947). The river will normally come up quite slowly over a period of several days as the runoff from the upstream areas reaches the floodplain. Examples include the Severn and Thames.

Flash flooding will occur in small steep catchments and create short term events lasting only hours and sometimes only minutes. Examples include the Lynmouth disaster and the 2004 Boscastle flood. Many of these catchments now have an emergency warning system which gives advance notice of an impending flash flood.

Between these latter two are the **minor rivers** which tend to react to rainfall in hours rather than days but have some floodplain so do not cause flash flooding. The nature of the drained catchment will determine the type of response which the river has to rainfall as developed areas respond much more quickly than rural catchments. Prime examples would include the Tame, which drains Birmingham and much of the Black Country plus the rivers Test and Itchen.

Dam Burst. This is a possibility, where a water retaining structure, such a dam or canal, fails releasing water downstream. The effects are unpredictable and can be catastrophic but are thankfully rare.

Highway drainage can be a source of flooding especially during short intense summer storms and can be seriously affected by the runoff from fields.

Sewer flooding is usually caused by overloading of the system by rainfall or by malfunction due to blockages which may only manifest themselves during heavy rainfall. Inadequate capacity in a pumped system or mechanical breakdown can also be a cause.

Advance Measures

Most advance measures involve civil engineering and capital expenditure. Flood banks have been around for many years but, more recently, temporary barriers have become more common in urban areas. The protection of individual properties comes in two distinct flavours: hard measures which physically help to prevent flooding or mitigate the effects and soft measures concerned with managing the situation. There is now much advice on these measures available on-line and there are specialist companies who will carry out a survey.

Dealing with an Event

Once heavy rain has started to fall, knowledge of when and where flooding will occur is crucial to management of the situation. This is where some knowledge of the catchment characteristics is useful. Do flood waters take three minutes, three hours or three days to come down? If the area has flooded before, then a Flood Warden should have been appointed. Assuming that you have planned properly then the event itself should be manageable but mark maximum flood levels on the wall of each affected room and take photographs. Notify insurers and keep copies of all correspondence and receipts. There remains the unenviable task of mopping up.

Roles of the Various Parties Concerned with Flood Management

National Government and Treasury	• Policy and funding • Coordination with military
County Council	• Emergency planning and support • High level development planning
Police	• Security • mergency control and coordination
Fire and Ambulance plus Mountain Rescue	• Rescue and support
Environment Agency	• Permanent and temporary measures to mitigate flooding • Flood measurement and prediction • Maintenance of 'main rivers' • Training Flood Wardens
District Councils	• Local support such as provision of duckboards and sandbags • Support during 'mopping-up' and drying out • Provision of temporary evacuation facilities and buildings • Permissive powers to maintain 'non-main rivers' • Local development planning and building regulations
Parish Councils	• Provision of flood wardens in association with the EA • Volunteers
Drainage Boards	• Provision and maintenance of waterways under their control
Water Companies	• Provision of water supply and sewerage but not flood related measures
Volunteers	• Manpower for temporary measures • Provision of foodstuffs in the event of prolonged events
Insurance companies	• Reimbursement of insured costs

Note that the water companies have no substantive involvement as their responsibility is to supply water and take sewage away. The Environment Agency now has those powers and responsibilities which were held by the water authorities/companies before privatisation.

Frequency

The expression "a one-hundred-year-flood" is widely misunderstood but it did, once upon a time, mean a flood that will occur once in a hundred years <u>on average</u>. Thus, there is nothing to prevent two occurring in successive years. With the paving of urban environments and global warming, the frequency of extreme events has increased whether we like it or not. These are the years when floods occurred which would have affected my sister's house at Diglis in Worcester as recorded by the Water Gate of Worcester Cathedral:

1672, 1795, 1852, 1886, 1924, 1940, 1947, 1948, 1960, 1961, 1965, 1968, 1990, 1998, 1999, 2000, 2002, 2007, 2014.

Case Study - The Somerset Levels, 2014

I chose the 2014 flooding of the Somerset Levels as it was very newsworthy and largely misunderstood by many of the professionals even as much as the media and public. Following heavy rainfall on saturated ground, virtually all of the Levels below the 5m contour flooded for about a month. Pumping at Burrowbridge came to naught and the locals all blamed the lack of dredging in the River Parrett. Examination of a simplified diagram tells the true story.

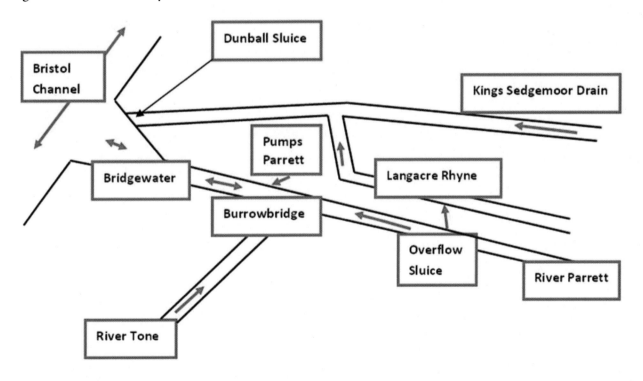

Note that all of the minor infrastructure such as sluices and pump stations have been omitted from the diagram as they did not affect the overall situation. The complications which need to be taken into account:

- The River Parrett is tidal back as far as Burrowbridge
- The lowest point of the whole system is at Dunball Sluice on the Parrett Estuary
- This is closed whenever the tide in the Bristol Channel is higher than the level in The KSD
- The Kings Sedgemoor Drain is the major outlet, not the tidal River Parrett

When the Levels flooded there was an outcry and a media storm soon resulted. Whilst there was flooding elsewhere in the country, they choose to concentrate on the Levels as it made better copy even though Bridgewater was not affected. The locals all blamed the Environment Agency who had ceased dredging the tidal reach of the Parrett as it was of no consequence. If a tidal river has twice the capacity when flowing out, it will bring twice as much back in with the tide.

Strangely, the locals, who had worked the Levels for centuries, did not appear to understand what was happening and neither did the EA at first. Temporary pumps were installed in the flooded area near Burrowbridge and they could be seen nightly on the television news, showering their outflow into the Parrett. This, however simply helped to back up the river which was already overflowing to the Langacre Rhyne upstream and hence to the KSD. Eventually, the penny dropped with the EA who asked the Dutch what they would do and then imported some rather large pumps from Holland. When these were installed at Dunball, the problem was soon solved.

There's a lesson in this as every flooding situation is different and it requires a solution (if one is available) that is tailored to the locality and the risk. Contrast the Levels with the steep valley through Ironbridge where barriers have been used to hold the River Severn back.

Paving your driveway

Flooding is somewhat unique as when it affects your property, it always someone else's fault and you want them to put it right at their expense. If you have a nice view over the river on a serene summer's evening, just remember that all that rain coming down from Wales has to go past your house.

And there's another thing which we are all guilty of – we paved over our drives. When we bought the house, and the kids were small, we only had one car but they grew up and now we have four cars. So we paved over the front garden so that we could keep them safely under our watchful eye. When the local surface water drains were designed, this was never envisaged but, thankfully, most surface water sewers have a good factor of safely. However, the cumulative effect of all this concentrated run-off is that your neighbours downstream now flood.

And that's not taking the change in weather patterns into account. The storms themselves are changing and getting more intense plus periods of wet weather are getting longer.

Main River

The expression 'main river' has legal, responsibility and practical connotations as it describes the watercourses which the EA has to maintain. 'Main rivers' are usually larger rivers and streams; they are designated as such and are shown on the 'main river map'. The EA carries out the maintenance, improvement or construction work and manages flood risk. Other rivers are called 'ordinary watercourses' where lead local flood authorities, district councils and internal drainage boards carry out flood risk management work. The EA have responsibility for any changes to a river's designation and the criteria for main river are primarily directed at the management of flood risk:

- A watercourse should be a main river if significant numbers of people and/or properties are liable to flood. This also includes areas where there are vulnerable groups and areas where flooding can occur with limited time for warnings

- A watercourse should be a main river where it could contribute to extensive flooding across a catchment

- A watercourse should be a main river if it is required to reduce flood risk elsewhere or provide capacity for water flowing from, for example, a reservoir, sewage treatment works or another river

- Short stretches of watercourses should be avoided i.e. alternating main river and ordinary watercourse to avoid multiple authorities acting on the same watercourse

- In the case of other watercourses, those responsible should have sufficient competence, capability and/or resources for flood risk management, including whether their governance enables sufficient competence, capability and/or resources, and local accountability.

Dredging and Navigation

Dredging may be carried out to improve the flow in a watercourse or as an aid to navigation. This duality leads to much confusion amongst the general population when flooding occurs as they naturally assume that it is carried out to avoid a river 'bursting its banks.' This is actually quite normal and helps alleviate downstream floods as the excess flow spreads out across the flood plain where it is effectively stored until the peak flow passes and the water subsides.

However, most of our substantial rivers have been adapted for navigation with the installation of weirs and locks along their length. Where this is the case, dredging is carried out purely for the purpose of navigation as the depth of the channel in now irrelevant - the carry-forward flow in the river is entirely dependent on the characteristics of the weir.

Without getting too deep (excuse the pun) into hydraulics, the flow in a watercourse depends on its gradient as well as the cross section. When unrestrained, the gradient is simply the angle of the bed of the stream or river. However, when we alter it to aid navigation, then the 'hydraulic gradient' becomes the parameter which determines how it flows. This is the slope that the <u>surface</u> of the water shows and is totally dependent upon the characteristic of the weir – not the bed of the river.

The reason for dredging a tidal reach, such as the River Parrett in the Somerset Levels, is less obvious as doubling the capacity of the outflow, at low tide, also doubles the inflow at high tide (did I already say that?).

Drought

At school, many years ago, we were taught that the definition of drought in the UK was 'no rain for more than two weeks' or words to that effect. This seems, quite rightly, to have withered on the vine and we don't seem, now, to have a definitive form of words.

According to Wikipedia, droughts are a relatively common feature of the weather in the United Kingdom, with one around every 5–10 years on average. These droughts are usually during the summer, when blocking high pressure (a cyclone) causes hot, dry weather for an extended period.

Anyone watching the weather forecast on a regular basis will soon become aware that the UK does not have consistent rainfall across the country and that Northern Ireland and the west coast of Scotland have very much more than other areas.

It is also clear that, whilst much of the country is self sufficient, the South East and the capital have very much less underlying resource. This means that the long-term planning for those areas needs to be more stringent than for areas where rainfall is abundant. This does not mean that those areas in surplus do not suffer droughts, rather that they are less frequent and less severe.

When a water company is suffering a severe shortage, they may request a 'drought order' which gives them powers to restrict supplies to customers.

Our main defence against drought is to make the basic resources more reliable. We achieve this through a variety of measures such as constructing reservoirs, raising the water level in lakes and supporting river flows so that abstraction is not affected. As droughts often affect a relatively small area, some water companies have constructed a 'strategic grid' which connects their main treatment plants so that they can support eachother.

All water companies are bound to supply a certain amount of water to customers even during times of drought. This can be achieved through restrictions such as cutting off the domestic supply and only allowing access via 'standpipes' which are installed at hydrants. Alternatives include supplying bottled water and the use of 'bowsers', which are trailers fitted with tanks which are filled with water.

Other Waterways

The Norfolk Broads

Probably the best known of the 'other' waterways are the Norfolk Broads which are now a National Park. It's said that this network of tidal rivers and shallow lakes are not natural having been made in the course of peat digging in the middle ages. OK, there may be proof for that but I wouldn't want to have been cutting peat in the middle of Hickling Broad during a wet winter. According to Wikipedia, there are seven rivers and 63 broads, mostly less than 4 metres (13 ft) deep. Thirteen broads are generally open to navigation, with a further three having navigable channels. Some broads have navigation restrictions imposed on them in autumn and winter, although the legality of the restrictions is questionable.

Hunsett Mill on the River Ant by Renata Edge

Like the Somerset Levels, much of the land is below the level of the rivers which drain it so some system, other than gravity, is required to get the water away. What most people think are 'windmills' are actually 'wind pumps' and these were the original means of pumping water from the lower level up into the tidal river which takes it to the sea.

The Broads have no locks on them and virtually the only restriction is Potter Higham Bridge, which has restricted headroom. This makes them popular for all kinds of boating whether sailing, canoeing or cruising in a houseboat.

On the Norfolk Broads by MixPix

The Fens

The Fens or Fenlands are the lowest lying area in the UK and much of the area is valuable agricultural land. Like the Broads, the water has to be pumped out and the original system used wind powered pumps. They comprise the low lying land which drains out though the Wash and extend from Lincoln in the north to Ely in the south. They are sometimes referred to as 'The Holy Land of the English' due to the number of monasteries in the area. They were also the home of Hereward the Wake who opposed the Norman occupation of England.

Wicken Fen with the wind pump

Canals and Navigation

A Bit of History

Man has used rivers for navigation for millennia so the principle is not new. What is new, though, is constructing a channel for the purpose of navigation. There are many examples of canals built in ancient civilisations but our concentration is on those in Britain where the Romans adapted waterways to aid navigation. According to Wikipedia, the Glastonbury Canal is believed to be the first post-Roman canal, built in the middle of the 10th century to link the River Brue with Glastonbury Abbey, a distance of 1,900 yards.

As a result of the industrial revolution, Britain was the first to develop a nationwide network of canals. The Sankey Canal of 1757, in the North West of England, was the first example which was followed in 1761 by The Bridgewater Canal between Worsley and Manchester for the transport of coal. Engineers like James Brindley became household names and the 'Golden Age' of the canals ran from the 1770s to the 1830s. In 1772, The Staffordshire and Worcestershire Canal was opened and in 1766, Josiah Wedgewood cut the first sod for The Trent and Mersey Canal (does what it says on the tin) which was opened in 1777. The aim was to transport raw materials into the industrial areas which needed them and then take the finished goods out for the domestic market and even for export via Liverpool and Bristol.

What followed was a network of navigable waterways, criss-crossing the country which survived until the railways took over the burden of moving goods around the country. Whilst the canals lost their business and many fell into disuse, their resurgence has depended on recreational use and the development of marinas as places to live. Remember, when next walking a towpath, that the barges were all drawn by horses in the first canal age. Imagine the chaos in Gas Street Basin as they tried to get past eachother on the single towpath. This issue was only solved when they started to use steam engines to power the boats and the 4 mph speed limit introduced. If you want more, Wikipedia has volumes of information and details of every canal in the country.

Channels

The early days of the canal building age were somewhat complicated as there were no standardised contour maps of the country. This meant that the great canal engineers such as James Brindley, had to do their own surveys. This was somewhat simplified as canals tend to follow minor watercourses for instance The Staffs and Worc's follows the Stour valley and the Birmingham Fazeley follows the River Tame.

In the early days, the canals tended to be narrow and were a little like country lanes with frequent locks. Later versions were wider - two boats wide and had fewer locks. Just compare The Stratford Canal with The Grand Union which was the M1 of its day. Narrow locks are just one boat wide whilst broad locks are two wide. The last major canal in England was the Manchester Ship Canal which is still able to support ships of up to 10,000 tons.

Most of the early channels were hand dug by 'navvies' who had to move around two cubic metres of earth a day (weighing about 3 tons, depending on the composition of the soil) and were paid on that basis. After the channel was dug, and shaped, it was made watertight by the use of 'puddle clay' which was 'puddled' manually to form the base of the channel.

The canal channels do not keep their water in them by accident. Many local streams are inter-connected with the canal and there are regular overflows as part of their quite sophisticated design. The 'summit pound' of every canal is problematical as it has to get its water from somewhere near the high point of the route. Whilst pumping has sometimes been employed, they are mostly fed by a reservoir via a 'canal feeder'. So many of the old reservoirs that you see around the country, are actually part of the canal network and have no connection with water supply.

Locks

Locks were standardised very early on and single locks are still 7ft wide and 72 feet long. Double locks, which can accommodate two boats at the same time are 14 foot wide. River locks are considerably bigger and their size depends on their traffic. The Caledonian Canal, which was completed by Thomas Telford in 1832, has locks which are 35 feet by 150 feet and takes ships of several thousand tons. We haven't used metric equivalents of the dimensions in this chapter as they would have been complete anathema in Napoleonic times when many of the canals were built.

Lock keeper's house on the Stratford Canal by Jlem6

Locks enable boats ('narrow' boats not 'long' boats – those were used by the Vikings) to traverse from a lower to a higher level and vice versa. The principle is simple as you 'lock' a short stretch of the waterway with gates which are opened and closed to allow the boat to enter and leave. Whilst in the lock, 'paddles' are opened to allow water to enter or leave as desired and the boat simply floats from one level to the other.

Single locks are the simplest and take just one boat at a time. They will usually have a stretch of canal between them in order to ensure there is enough water for them to operate but, if there is a steep hill to navigate, then they may employ a 'staircase'. This system of locks has no intermediate 'pound' and so the boat goes straight from one lock to another. In order to ensure that there is enough water, there will usually be side pounds which complicates the system of paddles.

Sometimes, there are two single locks in parallel (such as at Hilmorton on the Oxford Canal) and these are called 'tandem' locks. 'Double' locks are simply twice as wide so that they can accommodate two boats at the same time.

A lock 'flight' is a series of locks in proximity. Probably the best known are The Tardibigge Lift which enables boats on the Worcester and Birmingham Canal to get up to the Midlands plateau and the Caen Hill Locks on the Kennett and Avon. It can take more than a day to traverse its 30 locks.

Water saving or shallow locks are somewhat mystifying as there appears to be no obvious reason for them. There's one just outside Coventry where the Oxford joins the Coventry Canal and another at the junction of the Shropshire Union to the Staffs and Worc's. Their purpose is to avoid a younger canal stealing water from an older one. Even a six inch difference in the water level ensures that the flow will be from the later construction to the one which already existed.

Caen Hill locks on the Kennet and Avon Canal by Arpingstone

1 – open upper paddle and fill lock, close paddle

2 – open upper gate, boat enters lock, close upper gate

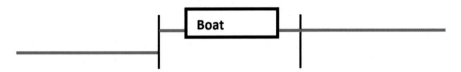

3 – open lower paddle and lower water, close paddle

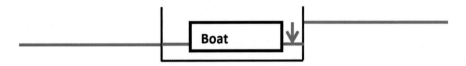

4 – open lower gate and exit boat, close gate

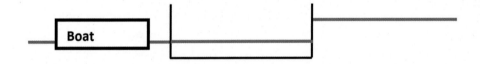

Basins and Marinas

Canals don't exist in isolation – they come from somewhere and go somewhere and with a purpose. Most canals were constructed to facilitate the movement of goods and so, one of their ends will be close to where the goods originate and the other end will be where they are intended to go. At each of these will be some sort of terminal in the form of a wharf or basin. Port Vale has a football team named after it and Wigan Pier was made famous by George Orwell. Of late, as recreation has become the main function of the canals, we usually call the latter 'marinas'. Of course, the highlight of late has been the conversion of Salford Docks (the terminus of the Manchester Ship Canal) into Salford Quays complete with Media City and the Lowry Theatre.

Media City by Martin Sylvester

Case Study – Stourport

Whilst the upper end of the Staffs and Worc's was spread out to serve industry in the Black Country, the southern end is quite unique and has a fascinating history. When Brindley first planned the canal route, he wanted the terminus on the River Severn to be at Bewdley which already had a dockside to serve the Severn Trows which brought stuff up from Bristol and took goods down for export. The trows, which were too wide for a canal anyway, were sailed up the navigable river and hauled up through rapids by men and horses.

Severn Trow at Blist Hills Museum by Jerrye and Roy Klotz

But the residents of Bewdley, who owned the existing dock, said that they didn't want his "stinking ditch" so he went three miles downstream to build his terminal basin. Thus the town of Stourport-on-Severn sprang up and is still a good example of Georgian architecture if a little seedy along the main street now.

Stourport Basin by Peter Evans

Winding Pools

Have you ever thought about turning a 70ft boat around in a canal which is less than 20 foot wide? Obviously you can do this in a basin or marina but elsewhere it's impossible unless you have a 'winding pool'. This is a short section of canal which is widened to allow a boat to turn through 180 degrees using the wind to turn it. There are not very many of them so journeys need to be carefully planned. Note the pronunciation is "winding" as with 'wind in the willows' not "winding" as in 'winding a clock'.

Winding pool by Andy Beechcroft

Canal and River Trust

British Waterways managed the canals and navigation on the inland rivers. The 1960s saw the formation of the Inland Waterways Association which was set up by enthusiasts who wanted to ensure that the canals of Britain, which were falling into disuse, were looked after for posterity.

In 2012, the government decided that it did not want the financial burden of managing rivers and canals and so set up The Canal and River Trust to manage navigable waterways. However, they left the responsibility for water quality with the Environment Agency as well as responsibility for non-navigable main rivers.

Tug 2 by Chris Morris

Recreation

Reservoirs

There is an ever increasing demand for potable water so, assuming that enough source water is available, the solution is usually to increase raw water storage by constructing a new reservoir. Such proposals cause a negative reaction, especially amongst people who live near the proposed site of the reservoir. Even with government and planning permission, there is significant opposition from many quarters including environmentalists, bird protection organisations, wild flower specialists, animal rights groups, etc. and, of course, the people directly affected.

When Anglian Water built Rutland Water in 1978, two small hamlets, Nether Hambleton and Middle Hambleton were demolished and flooded and the handful of residents re-homed in nearby villages. Severn Trent Water partially chose the site for a new dam at Carsington because only two farms and eleven people were affected. A protest group had been formed to oppose the dam construction but once the site was up and running in 1991 the locals all accepted the benefits that the new facility brought to the area. This is a common story, in spite of the disruption to the lives of a small number, protests are widespread but, once the reservoir is completed and full of water, the attitude of most people changes. Once there was simply countryside with the occasional house or farm but now there exists an amenity that can be enjoyed by a great number with consequential economic benefits.

The presence of water attracts recreational pursuits including: water sports, walking, cycling, fishing, bird watching or simply enjoying the flora and fauna that can be observed in the nature reserves around the site. Many sites offer sailing or windsurfing lessons, bicycle hire and bird watching. The track and paths around a reservoir are generally fairly flat and, more often than not, there will be a cafe and viewing area which makes the site ideal for people of all ages and even the less mobile. Many people sail and bring their own dingy to the reservoir whilst other will rent a boat for a few hours.

Water companies provide details of the reservoirs that can be visited and what amenities are on offer. Most have adequate parking facilities.

Angling

It was once said, that angling was the most popular 'sport' in the UK but it seems that numbers may have dropped off a little since then. Most regular anglers will have membership of a club and a choice of sites to fish which may include lakes, streams, rivers, canals, reservoirs and commercial ponds. Many specialise in coarse fish or salmon fishing and have appropriate gear for their hobby. You nearly always need a club membership, licence or day-ticket. The EA sell day-tickets for many of their waters.

Boating

Many of our inland waters support boating and the canals have a permanent population of those who choose to live on narrow boats, Areas like the Norfolk Broads have built an industry on it and a short trip along the Thames, upstream of the capital will tell you how popular it is. What would it be like without Henley?

Punting on the Cam by Timo_w2s

Canoeing

This is the sedate version of boating in a single or double canoe which can be like the tradition ethnic Amerindian version or just a sedate version of the kayak.

Canoeing for Waterchestnuts by US Fish and Wildlife Service

Kayaking

Kayaking can be a way of getting around but is generally more suited to the adrenaline junkies who seek out 'white-water' for their thrills. It's also a major sporting event and special courses have been built such as the one at Holme Pierpoint on the River Trent near Nottingham. It's also an Olympic Sport.

Kayak by Be creator

Rowing by Cheetah

Rowing

This is probably the most competitive water sport and many towns and cities have rowing clubs with boat houses alongside the river. Regular events are held throughout the summer and the best known is the annual Boat Race on the Thames between Oxford and Cambridge Universities.

Sailing

Sailing in inland waters is restricted by the size of the water that you are sailing on. Whilst some see this as a restriction, others think that it adds to the skill requirements as you have to get to know the wind as well as the water.

Learning to sail by 49,000 photos

Swimming

Many try to restrict swimming in open waters, especially flooded quarries where the water can be deep and get very cold. There are many groups who prefer open water to the local baths and even go out during very cold weather.

Photo by Seattle Parks and Recreation

Walking and Cycling

Canals and reservoirs tend to have quite gentle paths and tracks around them (the Elan Reservoirs excepted) as the water which is enclosed has to be all at the same level. This suits many who like to walk or cycle around or along them.

Photo by Simon Johnston

Bird watching

These days most of us just want to watch the birds or even tick them off on our list of those we have seen. Are you a 'dude' or a 'twitcher'? The former has had a good day out if he's seen a heron whist the latter knows all the birds in the book and even carries it with him alongside his binoculars or telescope.

Bird watching across the Refuge by USFWS Pacific

Wild-fowling

Water facilities attract wildlife and especially birds including lots of wildfowl. In days gone by there were many versions of wild-fowling but the use of a duck decoy, complete with a dog masquerading as a fox, must take the biscuit.

Punt gun by Jayfreedem

Bog Snorkelling

Why would anyone want to do that?

Rud-gr

Overall, water provides such varied environments from the mundane and serene to the downright dangerous – everyone to his/her own taste.

Utilities

As a 'utility' itself the water industry has much in common with those other organisations and businesses which provide the basic needs of modern society. We all have similar management structures, finance, plant and pipelines/cables which allow us to serve our customers whether we are part of (local) government or privatised. But the main source of interaction (other than paying the electricity bill) comes when we want to dig up the street to repair a pipeline or construct a new one. At a conference on utilities, the opening speaker commented that there is often more 'traffic' running <u>under</u> the street than <u>on</u> it.

Water

We have detailed our water and sewage pipelines elsewhere in this book so will not go into further detail other than to remark on their position in the street. Water distribution pipelines are usually fairly shallow and laid in the footpath whereas sewers are much deeper and tend to follow the centre line of a street or highway. They are (almost) never overhead.

Telecoms and Cable

Telecoms cables are laid in ducts when they are underground and tend to be strung between poles when overhead.

Gas

Traditionally gas pipes were made for cast iron but the conversion to natural gas in the 1960s created problems with the jointing and now most gas mains are made of plastic – usually HDPE. Gas distribution pipes are normally laid in the footpath and coloured yellow.

Electricity

Electricity cable may be ducted or wrapped in insulation and many are strung overhead between pylons – especially the higher voltages. Local, low-voltage cables are normally laid in the footpath but away from the gas main.

Oil

The oil industry has relatively few pipelines as its distribution is normally by tanker. It does, however, have a few very high pressure pipelines which, based on the need to serve airfields in WW2, connect the Midlands with Avonmouth, Merseyside and the Thames Estuary.

Solid Waste

The movement of solid waste is generally by specialist collection vehicles and so there is no permanent presence in the highway.

Highways

The highway authority has the unenviable task of coordinating street works under The New Roads and Street Works Act, 1961 which replaced a number of earlier enactments.

NJUG and Street Works UK

The National Joint Utilities Group was set up to co-ordinate the way that utilities carry out their works. It was very active in the early 1980s when computers made the interchange of map based information easier. One of the problems, at that time, was that the Ordnance Survey did not keep up with new developments so NJUG produced its own specification for updating the base maps. At that time it also produced recommendations for the colours to be used on composite maps and also for the dyeing of plastic pipelines and ducting:

- Potable water – blue

- Sewers – black or brown

- Cable – green (but purple seems to be preferred)

- Gas – yellow

- Electricity – red

- Telecoms - black

- Oil - black

Amongst a number of published guidelines the preferred position of pipelines and cables in streets is indicated in diagrams along with the preferred order of laying in the roads serving new estates:

- Streetworks UK - Guidelines on the Positioning and Colour Coding of Underground Utilities' Apparatus
- NJUG - Guidelines on the Positioning of Underground Utilities Apparatus for New Development Sites

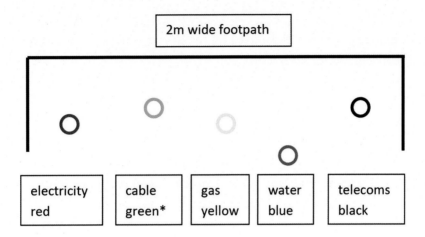

*whilst the standard says "green" most cable ducts now appear to be purple.

The sewer is normally in the centre of the street and much deeper.

The Drinks Industry

Whilst the UK water industry is big, it is dwarfed by the drinks industry which depends, to a large extent, on the provision of clean water as its base.

Bottled Water

Many do not like the taste or smell of tap water, often due to the presence of chlorine and fluoride and so prefer to buy bottled water at many times the cost of that supplied through the pipes of the water company. Some water companies sell bottled water and all of them have reserve supplies which they can use when the mains supply is unavailable. The majority of suppliers depend, to a large extent, on marketing the purity of their product which depends to some degree on its source. The Alps and local springs are favourites.

Tap water costs around £1.50 per m3 which is 0.15p a litre whereas supermarket bottled water retails at between 20p and 80p per litre – but then they have the cost of the plastic bottle.

Soft Drinks

By 'soft' drinks, we are referring to non-alcoholic drinks supplied in bottles. They all rely heavily on added flavourings which are often based on secret formulas and come either as 'still' or fizzy (carbonated). Having passed stringent safety criteria, they are often relied upon when the local tap water is not considered to be safe.

Beers

The importance of alcoholic drinks through history should not be underestimated as weak beer was the staple drink of choice through the middle ages. This was because boiling of the water before brewing made the water clear of parasites and hence safe to drink. Addiction to alcohol is another reason why such drinks are popular if not essential in most modern societies.

Beer is made from barley which has been malted to aid the development of sugars in the 'worts'. When yeast is added to the mix, the sugars are turned to alcohol at around 2-5% strength. Burton-upon-Trent became the brewing capital of England and this was largely due to its location on an aquifer which is rich in sulphides. Sulphides make brewing easier and result in a tastier beer – so many say. The addition of hops adds to the flavouring and makes the beer last longer. India Pale Ale (IPA) relied on a good dollop of hops in the brewing process to enable the beer to last all the way, by ship, to India for the troops serving there.

Whilst most towns and cities once had their own local brewery, now most ales are contract brewed in establishments which resemble factories as much as breweries.

Wines and Fortified Wines

The wine making process is somewhat similar but the sugar content in the grapes is much higher and so the alcohol strength is greater – typically 8-13%. Fortified wines, such as sherry and port, benefit from the addition of a distilled spirit and they usually have stronger flavours.

Spirits

Having made a weak alcohol, many like to make it stronger and the best way to do this is by distillation.

Scotland has made a whole industry out of it as has the state of Kentucky in the USA. Most countries have their own version of spirit.

You can make a basic alcohol out of almost any plant based material but whisky is based on distilling the basic alcohol which, like beer, is made from barley. Grapes (for brandy), grain and even potatoes can be used. The base alcoholic stock, once distilled, is clear and (despite whatever they tell you) virtually tasteless like vodka. Most of the taste comes later; for instance with whisky, it comes from the sherry barrels that it is stored in. Anise is the favourite flavouring in many countries.

Hot Drinks

Tea and coffee are the most popular hot drinks but there are many others and some have medicinal properties. Virtually all hot drinks are water-based infusions of plant material which has often been dried in order to transport and market it.

Marketing

Whatever they tell you, your drinks are all water based solutions with different flavourings and varying degrees of alcoholic content - but don't let me spoil your favourite tipple,

This Sceptered Isle

"This royal throne of kings, this sceptered isle, This earth of majesty, this seat of Mars, This other Eden, demi-paradise, This fortress built by Nature for herself Against infection and the hand of war, This happy breed of men, this little world, This precious stone set in the silver sea, Which serves it in the office of a wall Or as a moat defensive to a house, Against the envy of less happier lands,--This blessed plot, this earth, this realm, this England."

Richard II by William Shakespeare

We have spent much of our time writing about various aspects of the 'inland' water industries in the UK but our place in history and the world depends very much on us being an 'Island Nation' and, as such, very dependent upon and familiar with – the sea. For coastal regions, salt water is just as important to them as their drinking water. For the sake of completeness we refer here to some of our industries which rely on the sea.

Naval Power

Britain, as a world power, largely relied upon its navy to protect its international trade, to expand its empire and protect itself from others. The ports associated with the navy, rely heavily upon it for their living especially Portsmouth and Plymouth.

HMS Belfast by Robert Pittman

Merchant Navy

The UK remains a major international trader and its merchant navy one of the largest though many British ships are registered overseas. London and Liverpool were, for many years, our largest ports but the introduction of containerisation has brought container ports like Felixstowe to the fore.

Cutty Sark off Falmouth by Andy Roberts

Deep Sea Fishing and Fish Farming

Fish and chips is considered to be our national dish and for good reason. The catch of white fish, especially cod, hake and halibut is the staple of the deep sea fleets but seasonal catches, especially herring are also important. Scotland sends most of their catch south from Peterhead and Aberdeen. Fish farming of salmon is now common especially in Scottish lochs.

Trawlers on way for night fishing by Sheba

Ship Building

With three navies to support (Royal, Merchant and passenger), the country had major ship-building businesses especially along the River Clyde and in Belfast. Whilst the industry is well past its heyday, many smaller ship builders remain though not on the same scale.

Titan in Glasgow by Paisleyorguk

Ferries

Look at your road map and see the dotted lines crossing the surrounding seas in all directions. Ferry trade and passenger movement is a major industry in its own right especially in those ports which are in proximity to our neighbours. Dover and Folkestone remain the shortest route to the continent and are, hence, our busiest ferry ports.

Dover Harbour Aerial by John D Fielding

North Sea Gas and Oil

The discovery of gas in the North Sea, revolutionised our power industry and had major economic benefits to the country. Aberdeen benefitted when it became the major centre for North Sea activities.

North Sea Oil Rig by Tips for Travellers

Wind Power

With the move away from fossil fuels, off-shore wind power has become an important contributor to the country's source of power.

Off-shore wind farm, Cromer, Norfolk by Orangeaurochs

Recreation

Of all the major industries which rely on the sea for their very existence, the seaside resort is still, perhaps, the one we are all most familiar with. A week or two (if you were lucky) was a staple diet for most of us in the industrial towns and cities until it was overtaken by our desire to fly to Benidorm. Many seaside towns have been in decline for many years - just look at Blackpool and Morecambe for proof.

In case you were wondering – yes that's me with Eric in Morecambe

If you want a guarantee that the water is safe – look for a 'Blue Flag' beach.

Blue Flag Beach 2009 by Aspaonline

The "Fourth Rescue Services"

No chapter like this would be complete without mention of The Coastguard and The RNLI. Whilst the former is an agent of Government, and hence funded by them, The Royal National lifeboat Institution is a charity, run largely by volunteers and dependent upon donations. Both are on watch 24/7 to ensure your safety around the coast of Britain and on its waters.

RNLI Inshore lifeboat training off Weston Super Mare by forkcandles

Recent Disasters

Well, actually not all 'disasters' but some of the newsworthy problems suffered by the industry over the years which indicate why you can never relax. I hesitated to write this chapter as it can be seen as somewhat negative but then, forewarned is forearmed. You might query why some of them are included here in a book about the water industry but the text should reveal a relevant link. Many are not connected with drowning.

1947 Floods

The winter of 1946/47 saw heavy snowfalls in Wales and the North which persisted until the following spring. When a deep depression followed, the heavy rain melted the deep snow which brought about severe flooding all along the Severn Valley. If you get to visit Worcester Cathedral, have a look at the flood marks on the wall by the Water Gate and note the size of the sturgeon outlined on the wall behind the gate.

Flood levels, Worcester by Philip Halling

1952 Lynmouth Flood

In August 1952, a storm of tropical intensity broke over SW England, depositing nine inches of rain on the saturated soil of Exmoor. There has been speculation that the intensity of the rain was caused by experimental cloud-seeding operations. Debris-laden floodwaters cascaded down the steep valley from the moor and converged on the village of Lynmouth. In the upper West Lyn valley, fallen trees and other debris formed a dam, which in due course gave way, sending a huge wave of water and debris down the river, the natural course of which had been altered. Thirty four people died and over a hundred buildings were damaged.

1953 East Coast Floods

Coronation year was not good on the East Coast and along the lower Trent Valley. High winds and low pressure drove a tidal surge down the North Sea coast which over whelmed coastal flood barriers. At more than five metres above normal high-tide, it surged up the Humber estuary and the lower reaches of the River Trent. Over three hundred people died in Lincolnshire, Norfolk, Suffolk and Essex and nineteen in Scotland as well as those who passed away in other countries bordering the North Sea. Huge areas of countryside were flooded and many small ships were lost at sea.

1962/63 Winter

The winter of 1962 into 1963 was very harsh and even caused a potato famine. Its effect on the water industry, beside interfering with normal operations, was mainly felt in the thaw which brought to light a large number of burst water mains. This resulted in formal guidance being issued which required water mains to be buried deeper, thus avoiding the effects of frost.

1966 Aberfan

You might question what part water played in the Aberfan Disaster which killed 116 children and 28 adults when a coal tip slid down the hillside overlooking the town and engulfed the primary school. Those involved in soil mechanics are always aware of the part played by water in lubricating tips and structures made from earth or mine waste. After heavy rain, the tip, which covered over a number of natural springs, slid down the hillside engulfing the school and part of the village. If those responsible for the tip had taken note of the warnings issued by locals, it should never have happened.

1975/76 Drought

I had just joined the industry from local government when this all kicked off. Having been through two major reorganisations, we were not prepared for a major drought though most of us survived with our tap water intact. The dry summer of 1975 combined with a dry winter meant that many reservoirs were already drained down. The '76 drought started in early May and there was not a drop of rain through to September; it was said to be the driest summer for 200 years. This resulted in restrictions such as hose pipe bans and pleas from all and sundry to save water. A colleague published an article asking householders to put a brick in their toilet tank and the Chief Executive assured everyone that urine was safe to leave in the pan. Denis Howell, the "Minister for Rain" visited but only seemed interested in his press coverage which he bolstered by 'announcing' a link between the East and West Midlands supply systems (this took over ten years to complete and has hardly ever been used). He also boasted of taking a bath with his wife and did a rain dance for TV. Later in 1976 he became "Minister for Floods" and in 1978 "Minister for Snow."

Thankfully we appear to have moved on.

1976 and 1985 Legionnaires' Disease

Legionella is a group of pathogenic bacteria which acquired its name after an outbreak of a "mystery disease" at a convention of US military veterans - the American Legion. It made 221 people ill and caused 34 deaths. It wasn't until the next year that the causative agent was identified as a previously unknown bacterium which was subsequently named Legionella. Numerous outbreaks have occurred since (e.g. Stafford District Hospital, 1985 resulting in 25 deaths) and they are usually associated with poorly maintained air conditioning systems which carry the bacterium in an aerosol which comes from the cooling system.

UV treatment of the AC system's water is very effective and most large cooling systems now have a 'water safety plan' which reduces the risk of transmission.

1983 Water Strike

In January 1983, during the Thatcher Years, 29,000 water workers (but not staff) went on strike for more pay but this did not become the predicted disaster. The Water Authorities were well prepared and used contractors to carry out essential work. After three weeks, the dispute was referred to a Special Commission which made a settlement between the parties.

The Miners' Strike was to come in the following year.

1984 Abbeystead

On the evening of 23 May 1984, 44 visitors were attending a public presentation by North West Water who were demonstrating the operation of the station. Methane had seeped from coal deposits far below and had built up in an empty pipeline. When the pumps were switched on, the gas was sucked into the valve house and was ignited causing an explosion. It was so severe that it caused the concrete roof to fall down on to the group, destroying the steel mesh floor and throwing some of the victims into the water chambers below which rapidly filled with river water. The cause of ignition has never been determined but could have been caused by a lighted cigarette.

The dangers from methane gas became better understood as a result and regular gas testing has since been carried out at susceptible locations.

1984 Carsington Dam Failure

Whilst there had been contractual issues with the tunnel construction, no-one foresaw the much larger one with the dam. It was designed by a relatively inexperienced consultant and had a very slim profile for an earth dam. Under the large strains induced, the foundation clay - for which there had been very limited strength testing - suffered local brittle failure, resulting in progressive failure of the dam's upstream face. It had to be dismantled and totally rebuilt with an improved design, overseen by an experienced consultant. Four later died at the site though not as a result of the dam failure.

1985 The Manchester Air Disaster

British Airtours Flight 328, en route to Corfu, was a tourist flight which was about to take off when an engine caught fire at Manchester Airport. All international airports have full-time fire-fighters on site and they attended the incident very quickly but were delayed in their attempts to control the fire by the lack of foam which is essential in dealing with aircraft fuel fires. It resulted in the loss of fifty five lives, mostly through smoke inhalation and led to new safety regulations on aircraft. Recommendations were also made in respect of the maintenance of fire control mains and the provision of foam on airfields.

1988 Camelford

The water supply to the town and surrounding area was contaminated when 20 tons of liquid aluminium sulphate was added to the wrong tank at the Lowermoor WTP on Bodmin Moor and hence got into the water supply. The long-term health effects on those that consumed the water remain unclear.

Intelligent key systems are now used on all such access points to avoid similar incidents.

1989 The Marchioness Sinking

The sinking of the Marchioness is a poignant reminder of the inherent dangers of venturing out onto water especially on a busy waterway and at night. Marchioness was a small pleasure cruiser which was run down on the Thames in the middle of London by a much larger bulk carrier, the Bowbelle, resulting in the loss of fifty one party-goers.

Most such craft now have on-board radar.

1994 The Worcester Incident (aka The Wem Incident)

Taken from 'A History of Worcester's Water Supply' by Joan Harris.

"On the evening of April 15 1994 I arrived back in Worcester from working in Birmingham, to be told that the City water supply was contaminated. The initial advice was that it should not be used for drinking, cooking or washing. The following day, instructions were that it could be used for washing and washing dishes, and by 17 April it could be used normally. Bowsers were brought into the city on the evening of 15 April, and people queued for water. Others, co-ordinated by local radio, ferried water to the aged and infirm.

The cause was a quantity of chemicals which had entered the river at Wem in Shropshire, and had only been detected when a number of users at Worcester complained of the taste and smell. All this brought the water supply (so often taken for granted) very much into people's minds. It gave me the incentive to write up what follows, based on earlier research. Subsequently Severn Trent Water gave each household a cheque for £25 for the inconvenience caused. I think that under the code of practice laid down in April 1636, there would have been no compensation, as this would have come into the category of "ill accident which the undertakers cannot prevent."

All hell broke loose as the company tried to move water around the region but it was completely unprepared for such an incident. South Staffs, who used the same source were largely unaffected as they closed their intake from the river at Hampton Loade and relied on their strategic network. ST had to rely on the Army to deliver and protect supplies in bowsers due to interference and vandalism by opportunist sellers of bottled water.

The perpetrator of the problem was later found innocent of charges brought against him and Severn Trent, as owners of the system, were found guilty and fined. Worcester's Barbourne water treatment plant was later closed and the City supplied by linked mains from Trimpley and Strensham. This then formed the basis of ST's strategic grid which had been proposed over a decade before. It still didn't protect The Mythe in 2007 (see later paragraph).

1995 Yorkshire Drought

Many areas in the Yorkshire Water supply area almost ran out of drinking water in 1995 due to a severe drought which brought their reservoirs to very low levels. There were serious threats of cuts to supplies but the company avoided this by the use of tankering from neighbouring areas – even Kielder Water. There was extensive coverage in the national press of water tankers queued up alongside the M62 motorway before discharging their loads into Scammonden Reservoir.

1998 Floods

Every flood has its own characteristics and we have written elsewhere on Lynmouth and Boscastle as well as a chapter on flooding which deals with the differences.

Around Easter 1998, the equivalent of a month's rain fell in the Midlands in 24 hours. Flooding caused over£400m of damage and there were five associated deaths. The Environment Agency commissioned a report which concluded that the EA's policies and operational arrangements were sound, but there were instances of unsatisfactory planning, inadequate warnings, incomplete defences and poor co-ordination with emergency services. The report also highlighted the need for flood warning systems and pointed out that the scale of the damage could have been avoided if more advice had been issued to those affected. People did not understand what they could do to protect themselves even when they are pre- warned.

As a result of the flooding experienced by my sister, who lived alongside the River Severn at Diglis, Worcester, I published my advice as *The Gospel According to Noah*. It's still available on WordPress.

2004 Boscastle Flood

89mm of rain fell in an hour over ground which was already saturated from previous rainfall. The topography of the land upstream of Boscastle comprises a steep-sided valley which acted as a funnel, directing a vast volume of water straight through the centre of the village. Scenes of cars being washed down were shown on national television.

As a result of the incident, which did not result in fatalities, new advance warning systems were installed and have since become commonplace especially upstream of vulnerable caravan sites.

2004 Morecambe Bay Disaster

A large group of Chinese immigrants were cockle picking in Morecambe Bay when they were cut off by the tide and at least twenty one died. Despite being warned by locals, they stayed on until it was too late and drowned. The gangmaster was sentenced to fourteen years imprisonment for manslaughter and perverting the course of justice.

The Queen's Guide to the Sands is an appointment made by The Duchy of Lancaster which dates from 1548 and has a salary of £15 a year. Michael Wilson was appointed in 2019 but Morecambe Bay remains a dangerous area despite warning systems having been improved.

2005 Fraud

In 2005, a major water company was fined over £36m for deliberately submitting false performance information to Ofwat. Whilst no-one died as a result, the effect on staff morale at the company was enormous. It was later found that another director has falsified the customer complaint figures and hence resigned.

2007 Floods

Despite being designed to withstand a '1947 flood', The Mythe Water Treatment Plant became inundated with flood water from the River Severn. The main pump house had to be switched off and the water coming into the plant was contaminated. This led to the loss of tap water for approximately 150,000 people in Cheltenham, Gloucester and Tewkesbury.

The Mythe and Strensham water treatment plants are now interlinked.

2017 Grenfell Tower Fire

Over seventy people died in the Grenfell Tower fire which burned, out-of-control up the outside of the block. Whilst all such blocks have a dry riser (see chapter on fire fighting), the system is not designed to deal with an external fire in the cladding. The reaction of the first fire crew to arrive on the scene indicated their complete inability to deal with such an event as it had not been anticipated. It had not been included in their training, nor were there any mitigation measures available to them. I won't go over the findings of the Commission but would ask some simple questions which relate to the way that fires are normally fought:

- would the situation been any better or worse if it was raining - we install interior sprinkler systems but why not exterior ones?
- why did the brigade not use the dry riser - was it because debris was falling off the block which prevented them from connecting it to the main?
- as water was available on the roof of the block, why was there no system in place to make use of it on the exterior? If it had been used, would it have given time for the whole block to be evacuated?

We don't see any sign of any of these questions being addressed at the enquiry as the whole concentration has been on apportioning blame and removing cladding.

Appendix 1 Abbreviations

For Bodies see Appendix 4 which covers them.

AM	Asset Management
AMP	Asset Management Plan
AMS	Asset Management System
AV	Air valve
BOD	Biochemical oxygen demand
Capex	Total capital expenditure
CCTV	Closed circuit television
CESWI	Civil Engineering Specification for the Water Industry
CG	Condition Grade
CIP	Capital Investment Programme
CMMS	Computerized Maintenance Management System
COD	Chemical oxygen demand
CSO	Combined storm overflow
DAF	Dissolved air flotation
DMA	District meter area
d/s	Downstream
DWF	Dry weather flow
GSS	Guaranteed Standards Scheme
LoS	Level of Service
LNC	Leak Noise Correlator
M&E	Mechanical and Electrical
NRW	Non-revenue water
NTU	Nephelometric Turbidity Units
NVQ	National Vocational Qualifications
OPA	Overall Performance Assessment
Opex	Total operational expenditure
PAM	*Principles of Asset Management*
PAS	Publicly Available Specification
PG	Performance Grade

PHA	Public Health Act
PRV	Pressure reducing valve
PSV	Pressure sustaining valve
RAG	Red, Amber, Green
RO	Reverse osmosis
RPS	Regulatory Position Statement
RWA	Regional Water Authority
SAMS	*Simplified Asset Management Systems*
SCADA	Supervisory Control and Data Acquisition
SIM	Service Incentive Mechanism
SS	Suspended solids
STC25	*Standing Technical Committee Report 25*
SUiAR	Sludge Use in Agriculture Regulations
TDS	Total dry solids
TS	Total solids
TOTEX	Total asset life expenditure
UFW	Unaccounted for water
u/s	Upstream
UV	Ultra violet
VIP	Ventilated Improved Pit latrine
WASH	Water, Sanitation and Hygiene
WASSP	The Wallingford Procedure
WO	Wash out
WSC	Water and Sewerage Company

Appendix 2 Glossary

The explanations given below are based on the specific meanings used in the water industry rather than those found in a dictionary. Beware of taking them too seriously as I made some of them up on the spot.

Adoption

A legal process whereby a water company takes ownership of a pipeline and responsibility for it

Advanced treatment

Treatment which uses high-tech means to treat water after the normal stages of treatment

Aqueduct

A pipeline or channel, constructed to carry (usually) raw water from one place to another

Aquifer

A natural body of water contained underground

Artesian

Applies to a natural system (a basin) where water is pressurised in its aquifer and naturally rises above the surface when allowed to escape

Asset management

This term has various meanings which include:

- <u>financial</u> asset management means looking after monetary assets including stocks and shares
- <u>infrastructure</u> asset management concerns the management of physical assets

Bacilus concretivorus

Concrete eating bacteria (does what it says on the tin!)

Basin

A low lying area of land which receives water and often has no outlet, relying on evaporation for water to escape

Bilge water

The foul waste water that collects at the bottom of a ship's hull

Black water – as opposed to 'grey water'

Sewage which containing faeces

Blue Flag beach

A beach which has been certified as being free of pollution and (presumably) safe to use for bathing.

Booster

A pump situated on a supply or distribution system which 'boosts' the pressure in the pipeline or network in order to serve an area at a higher elevation

Borehole

A vertical hole drilled through the ground to abstract groundwater from an aquifer

Bowser

A trailer which is filled with potable water and which is towed into position to supply those who have been cut off from the distribution system

Brackish

Describes water which has some salt in solution but not as much as sea water

Buoyancy

The principle that something which is less dense than water and will float on it

Canal

Based on the Latin (canalis) and Spanish for 'channel' a canal is an artificial i.e. constructed watercourse. Whilst internationally 'canal's may be used for the transfer of water from one place to another, its common usage refers to a navigable waterway.

Catchment

The area of land which drains to a particular watercourse

Close-coupled

Where a pump and its motor are connected by a very short shaft

Combined sewer

A sewer which has both foul and surface water in it

Combined storm overflow

A chamber on a sewage system which allows excess flow, in storm time, to overflow to an adjacent watercourse

Composting toilet

A toilet that works without water by composting urines and faeces

Consumer

One who buys goods or a service from a single supplier without choice

Culvert

A pipeline or other construction allowing water to pass under a structure such as a building or a highway

Customer

One who buys goods or a service from a supplier whom they have chosen

Dam

A structure to retain water

Desalination

The process of removing salt from (usually) seawater to make the water potable

Disinfection

The process of killing off harmful organisms

Distribution main

A small localised main which distributes potable water to properties along its route

Drain

This can mean almost any small pipeline taking water away but, more properly, it means a pipeline taking drainage from a single property

Drought order

A formal order, issued to a water company by government which enables it to restrict supplies

Easement or Wayleave

A legal entitlement enabling a pipeline or cable to cross private land

Effluent

A liquid which flows out from somewhere

Electrolysis

The process of separating hydrogen and oxygen gasses from water using electricity

Epidemiology

The science of determining outcomes from the examination of data concerning large populations

Faeces

Excrement, excreta, s**t, crap, cac, dung, turd, floater, poop, poo, droppings, manure, stool, ordure, etc.

Floodplain

That area of land in the middle reaches of a river which floods and, in doing so, alleviates the degree of flooding downstream

Force main and **rising main**

The pressurised pipeline leading away from a pump station. Whilst 'force main' is used internationally, the term 'rising main' is peculiar to the UK

Foul sewer

A sewer that contains only sewage and no rainwater

Grey water

Waste water emanating from premises but not containing toilet waste, i.e. faeces

Groundwater

(1) The water contained naturally underground which forms the basis for springs

(2) The water residing in an aquifer

Hardness

A measure of the concentration of calcium or magnesium carbonate dissolved in water

Highway drain

A pipeline which drains a highway

Hydrogen peroxide

A chemical consisting of two oxygen molecules and two of hydrogen; often used as a disinfectant

Ice

Water in its solid state (usually below 0° C)

Intake

The interface between a natural body of water and a treatment plant which processes it

Interceptor or interceptor sewer

A main sewer which intercepts minor sewers usually running in parallel to a watercourse

Lagoon

A constructed lake or pond

Lake

A natural body of non-saline water

Laminar and turbulent flow

Laminar flow is quiescent whilst turbulent flow is "what it says on the tin" i.e. turbulent

Latrine

A constructed pit, usually with some form of seating for the acceptance of faeces and urine

Lavatory

A washbasin/place to wash your hands; has come to be used instead of 'toilet'

Leakage

Potable water which physically escapes from supply and distribution networks

Main River

A legal definition applied to those watercourses which are the legal responsibility of the Environment Agency

Midden

A dunghill or refuse heap especially used for faeces and urine

Modelling

The building of a computer simulation of a process or network

Non-revenue-water and **Unaccounted-for-water**

Water which has been put into the potable supply system but which is not paid for by customers

Ozonation

A disinfection process using ozone in potable water treatment to kill pathogens

Package plant

A small treatment plant, so called because it is easy to transport, often as a single load on a lorry

Pasveer ditch

A proprietary system of sewage treatment

Pig

A device which can be propelled or pulled along inside a pipeline to separate liquids or clean the pipe

pH correction

Adding a substance to make water chemically neutral as an aid to treatment

POPs

Persistent Organic Pollutants

Preliminary treatment

A mechanical process such as screening prior to the main processes of water or sewage treatment

Primary treatment

Usually a mechanical process in water or sewage treatment to settle solids

Reverse osmosis

A technical process using very high pressures to clarify water by forcing it through a plastic membrane

Reservoir

A constructed lake which stores water for subsequent usage

Ring main

A particular kind of design for a supply or distribution system which connects back to itself thus allowing a degree of redundancy and pressure equalisation in the network

River

A large watercourse or one so named

Reed odourless earth closet

A chute, in conjunction with a ventilation stack, encourages vigorous air circulation down the toilet, thereby removing odours and discouraging flies

Sand filter

A common form of water treatment to take out small suspended solids; can be 'rapid gravity' or 'slow'

Secondary treatment

A biological water or sewage treatment process to improve oxygen content or remove contaminants

Septicity

The state of sewage which is starved of oxygen and goes black due to the creation of sulphides

Septic tank

A construction, consisting of one or more compartments to provide basic treatment to sewage, usually in rural areas

Sewage

Waste water emanating from domestic premises especially that containing urine and faeces

Sewer

A pipeline for the collection of sewage that serves more than one property

Sewerage

A collection of pipelines designed to accept and transfer sewage; may contain pumping equipment; Americans use the term as 'sewage'

Steam

Water in its gaseous state (usually above 100 degrees Celsius)

Stream or Brook

A minor watercourse

Submersible pump

A device which is designed to function in a liquid especially water:

(1) a close-coupled sewage pump which is installed in a wet well

(2) a combined motor and pump which is designed to abstract groundwater from a borehole

Tertiary treatment

A follow-on process to improve effluent quality through the removal of nitrates and phosphates in sewage treatment

Toilet or Lavatory

A device or cubicle used to accept faeces and urine to be taken away for disposal. Many words are used in place of it such as: rest room; the crapper; karzie; john; 100 (hence 'loo'); WC; etc. Actually the literal meaning of 'lavatory' is wash basin

Trunk main

A large water main which is transferring large quantities of water from one place to another rather than supplying the area through which it passes

Turbidity

The quality of being cloudy, opaque, or thick with suspended matter and its measurement in NTU

Ultra violet

UV is often used to kill any remaining organisms in treated water before chlorination

Utility

A business that provides one of the basic requirements for living; provided by government or other such public body

Village drain

An informal drain which takes drainage, often sewage, from rural properties, which provides a degree of basic treatment

(The) Wallingford Procedure

An early computerised sewer analysis program

Water

Compound consisting of one oxygen molecule and two of hydrogen; term normally refers to it in its liquid state

Water Authority

A public body which owned and managed water supplies and sewerage functions before privatisation

Water Company

A business which supplies potable water which may, or may not, also deal with waste water

Water course

Natural rivers, streams, brooks and other means by which water runs downhill

Water hammer

A banging noise produced by water pipes when they experience a sudden change in pressure

Water only Company

A water supply business which does not deal with sewerage functions

Watershed

The dividing line between two catchments which, in America, has been confused with 'catchment'

Water table

The surface level which groundwater has reached

Wet well and **Dry well**

A wet well is a constructed chamber which is designed to contain water whereas a dry well does not contain water. Many pump station have one of each. The wet well contains the liquid to be pumped and the dry well contains the pumps and they are separated by a wall.

'Yuk' factor

The reluctance of a population to consume treated effluent as potable water regardless of how well treated it is.

Appendix 3: Units

You might be asking yourself why would anyone need to write a chapter on 'units'. You already know that every industry has its own jargon – well, they also have their own set of pet units and unless you are familiar with them, you can't appreciate the magnitude of things.

Let's see an extreme example. America does not use metric units, which can have catastrophic consequences if you are trying to land a joint American/European mission on Mars but it gets even stranger than that – here's an extract from Wikipedia:

"The acre-foot is a non-SI unit of volume commonly used in the United States in reference to large-scale water resources, such as reservoirs, aqueducts, canals, sewer flow capacity, irrigation water, and river flows. An acre-foot equals approximately an eight-lane swimming pool, 82 ft long, 52 ft wide and 9.8 ft deep."

Clear? Note that they don't use metric units or say an "Olympic size swimming pool." Americans have also amended the spelling so they have meters and liters etc.

Even when we are talking about currency there are exceptions. The rupee is not worth very much on its own so India has a unique way of describing large amounts of currency: a 'lakh' is 100,000 and a 'crore' is 10,000.000.

The early 1970s, in the UK, were unusual because a number of important changes took place:

- Local government and water industry reorganisation
- Decimalization of the currency
- Metrication of the construction industry

So the complication of compiling a bill of quantities, in yards, feet and inches with pounds, shillings and pence, was at an end. Metric units and metric currency made everything easier. If Napoleon had appreciated that the British would never agree to learn French, then metrication might have come at least a century earlier.

We use the simple metric units of length depending on the subject – millimetres (mm), metres (m) and kilometres (km). For a pipe diameter we would use mm but for its length, metres and if it's a very long aqueduct then km. Area is square mm (mm2), square metres (m2) or hectares. Very large areas are measured in square kilometres (km2). Cubic units rely on the litre so we have millilitres - though most prefer cubic centimetres (cc) for small amounts then litres and then cubic metres (m3) for larger ones.

An industry which moves water around in great quantities relies heavily on flow rates so we use litres per second for relatively small flow rates and then cumecs (m3/s or cubic metres per second) for larger quantities. A derived quantity which is in frequent use is megalitres per day (Ml/d) which is a million litres in each 24hour period. An alternative, but numerically identical, unit to this is tm3/d or a thousand cubic metres a day – take you pick but I prefer Ml/d.

Consumption is measured in litres per capita per day (l/c/d).

I wrote, above, that the unit for pipe diameters is mm but that's not altogether true for two quite distinct reasons. When we look at pipeline records (sewers and water mains) we find that many of the pipes are still described in Imperial terms rater than metric. This is because their size, in inches, is a historical fact and can't be altered simply because we have gone metric. A 'six inch pipe' by any other name is still six inches diameter rather that 150mm and the term 'half-inch tap' is still used in languages other than English!

The second complication concerns the mode of manufacture of the pipes themselves. Pipes which are made using a mandrel (an internal mould) are specified by the diameter of the mandrel whilst extruded pipes, which are forced under pressure from an external former, bear its dimension. Spun pipes bear the dimension of their interior diameter.

Anyone delving into water distribution or sewers will soon become familiar with the common conversion factors:

4 inch	100mm
6 inch	150mm
9 inch	225mm
10 inch	250mm
12 inch	300mm
15 inch	375mm
18 inch	450mm
24 inch	600mm
30 inch	750mm
36 inch	900mm
42 inch	1050mm
48 inch	1200mm
54 inch	1350mm
60 inch	1500 mm

Generally, water mains get larger by 2" (50mm) steps and sewers by 3" (75mm) steps) until we get to about 30" (750mm) when they both get larger by at least 3" (75mm) intervals. The 'normal' minimum size for a water main is 100mm but for gravity sewers it is 'normally' 225mm. There are now many 6" (150mm) sewers which, until recently, were usually private sewers.

Those who look after the treatment and quality of water have their own set of preferred units. BOD and suspended solids are usually measured in mg/l (milligrams per litre) thus a 'Royal Commission' standard of 20/30 would mean that the BOD would be less than 20mg/l and the SS would need to be less than 30mg/l. For other trace pollutants we use ppm (parts per million) but the WHO uses µg/l for its standards which, if a litre weighs about 1,000g, then a µg /l would approximate to parts per billion.

Pressure is force divided by area. Using standard metric units, the textbook measure of force is the Newton so pressure is in N/m^2. This standard unit of pressure has been defined as the Pascal, where $1\ Pa = 1\ N/m^2$. This is not popular with those who manage water systems for good reason as pressure is fundamental to maintaining a good service. Some prefer to rely on the imperial units of psi (pounds per square inch) as this relates, for instance to the 30psi or so that we put into our car tyres and, unlike the metric system it's very practical.

Many highly pressurised pipelines use the 'bar' as a measure and this is related to weather forecasting. One 'bar' is equivalent to average air pressure on planet Earth which is about fourteen point something psi. Weather forecasters go on to divide it up into a thousand parts which they call millibars (mb) and this is what you see on the weather forecast. However this does not help with our water mains. Instead we generally use a system based on 'head'.

The term 'head' relates to the height that water would rise to in a vertical tube attached to a pipeline. Normal atmospheric pressure is here and there everywhere and hence forms a baseline but once we pressurise a pipe, we can measure that pressure, above atmospheric, in metres of water. This is both practical, easy to understand, easy to measure and normally stated as 'metres head'.

With the exception of the United States we measure temperature in degrees Celsius (°C). Once upon a time they were labeled degrees "centigrade" but this was changed as the word has another meaning associated with angles. The use of the Celsius scale in water technology is most convenient as water freezes at 0°C and boils at 100°C.

Turbidity is measured in NTU: Nephelometric Turbidity Units.

In the late 1970s, when the dereliction of underground assets came to the fore, especially with some spectacular road collapses in Manchester, David Young of North West Water came up with a practical way of describing the size of the hole – the DDB. The Double Decker Bus was (and still is) equivalent to about 75 m3. This in turn led to newsreaders, few of whom are noted for their performance in 'O' level maths, to come up with a whole alternative universe of unitary measurement. Some of the more interesting ones, most of which are still in common usage today, are tabled below:

Short form*	Description	Equivalence
Volume		
an OSP	Volume of an Olympic Swimming Pool	2,500 m3
a DDB	Volume of a Double Decker Bus	75 m3
Area		
a TC	Area of a Tennis Court (up to the fence)	543 m2
an FP	Area of a Football Pitch (professional)	7140 m2
a Cymru	Area the size of Wales	21,000 km2 approx
Height		
a Crouch	Height of Peter Crouch	2.0 m approx
a Nelcol	Height of Nelson's Column	52 m
a BT	Height of Blackpool Tower	Half an ET
an ET	Height of The Eifel Tower	Two BTs
Distance		
a hair's breadth	Width of a Human Hair	17-180 μm
a Mar	Length of a Marathon	26 miles (42.2 km)
a LEG	Lands' End to John O'Groats	600 miles (970 km)
a MaB	Distance Earth to Moon and back	770,000 km
a very long way	Distance Earth to Andromeda**	2.5 million light years
Time		
a Sunsec	Time taken for light to travel from the Sun	8.3 minutes
a Millenium	Time from the Norman Conquest to now	1,000 years
Other		
a NiH	Needle in a Haystack	I in 100 billion approx
a Brum	City the size of Birmingham	population 1 million

* We made these up on the spot, just to fill the table; hope you don't mind but we hoped, like a wet wipe in a poorly constructed 4 inch foul drain, that they might stick.

** for longer distances suggest you contact Professor Brian Cox, c/o University of Manchester, who is very knowledgeable on the subject.

Appendix 4: Bodies

Like the medical profession, sewerage engineers tend to bury their mistakes. Whilst many of these 'bodies' have been buried this was generally only after they passed their sell-by date.

ADA (Association of Drainage Authorities)

British Water (Trade body)

BTEC (Business and Technology Education Council)

C&G (City and Guilds)

CAT (Centre for Alternative Technology)

CCWater (Consumer Council for Water)

CRT (Canal and River Trust)

CIWEM (Chartered Institute of Water and Environmental Management)

Defra (Department for Environment, Food and Rural Affairs)

DWI (The Drinking Water Inspectorate, a department of Defra))

DWQR (Drinking Water Quality Regulator [of Scotland])

EA (Environment Agency)

Flood Re (Flood-reinsurance) joint insurance scheme

FoE (Friends of the Earth)

FWR (Foundation for Water Research)

HBF (House Builders Federation)

HSE (Health and Safety Executive)

IAM (Institute of Asset Management)

ICE (Institution of Civil Engineers)

IDB (Internal Drainage Board, of which there are many)

IEE (Institution of Electrical Engineers)

IMunE (Institution of Municipal Engineers) now defunct after merging with ICE

IMechE (Institution of Mechanical Engineers)

IoW (Institute of Water)

IWA (Inland Waterways Association)

MAFF (Ministry of Agriculture, Fisheries and Food) incorporated into Defra in 2001

NFBTE (National Federation of Building Trades Employers)

NIEA (Northern Ireland Environment Agency)

NJUG (National Joint Utilities Group) now appears to present itself as Street Works UK

NRA (National Rivers Authority) now defunct – part of the EA

NWC (National Water Council) now defunct

Ofwat (The Water Services Regulation Authority)

OS (Ordnance Survey)

PIG (The Pipeline Industries Guild)

RWA (Regional Water Authority)

SAS (Surfers Against Sewage)

SEPA (Scottish Environmental Protection Agency)

WaPUG (Wastewater Planning Users Group, now Urban Drainage Group and part of CIWEM)

WaterAid (charity)

WEDC (Water, Engineering and Development Centre)

WHO (World Health Organisation)

WICS (Water Industry Commission for Scotland)

WRAS (Water Regulations Advisory Scheme)

WRB (Water Resources Board) defunct since 1973

WRc (Water Research Centre)

Appendix 5: Management Structures

Management structures tend to be somewhat similar but never quite the same. Most successful ones are based, roughly, on the example which follows:

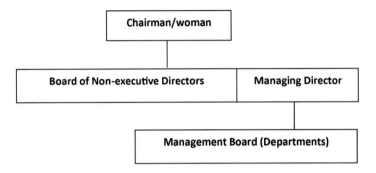

The departmental structures may look like this:

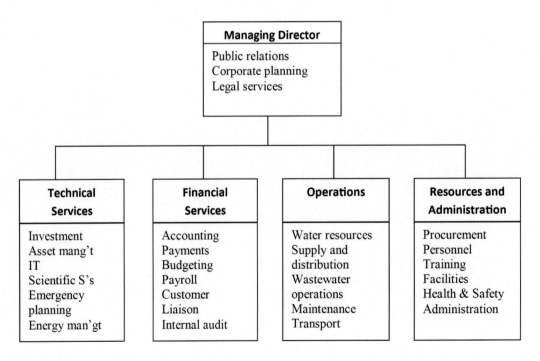

This structure presents a reasonably balanced set of functions between the departments. There is a constant argument within Technical Services and Operations as to how they should be divided up. One camp favours splitting them between capital and revenue functions whilst the other favours a split based on function i.e. between potable water and wastewater. This, inevitably gives an opportunity to any new manager to change from one to the other. In practice, this is unimportant as the quality of the managers is the key ingredient.

Appendix 6 – Regulation

Ofwat Performance Measures (aka 'levels of service')

These were introduced by Ofwat in 1990 and were used to compare the performance of the newly privatised water companies. The wording may vary according to the source.

- DG1 (adequacy of water resources or availability of supply)
- DG2 (risk of low water pressure or pressure of mains supply)
- DG3 (unplanned or supply interruptions)
- DG4 (restrictions on the use of water)
- DG5 (properties at risk of sewer flooding and flooding incidents)
- DG6 (response to billing contacts)
- DG7 (response to written complaints)
- DG8 (billing of metered customers or bills for metered customers)
- DG9 (telephone contact or ease of telephone contact)

Ofwat Key Indicators (since 2011)

The DGs were replaced in 2011 by a more complex series of measures.

Customer experience

Indicator	Definition	Measure
Service incentive mechanism (SIM)	The level of customer concern with company service and how well the company deals with them	Score
Internal sewer flooding	Number of incidents of internal sewer flooding for properties that have flooded within the last ten years	Number of incidents
Water supply interruptions	Number of hours lost due to water supply interruptions for three hours or longer, per property served	Hours per total properties served

Reliability and availability

Indicator	Definition	Measure
Serviceability water non-infrastructure	Assessment of the recent historical trend in serviceability to customers, as measured by movements in service and asset performance indicators.	Stable / Improving / Marginal / Deteriorating
Serviceability water infrastructure	Assessment of the recent historical trend in serviceability to customers, as measured by movements in service and asset performance indicators.	Stable / Improving / Marginal / Deteriorating
Serviceability sewerage non-infrastructure	Assessment of the recent historical trend in serviceability to customers, as measured by movements in service and asset performance indicators.	Stable / Improving / Marginal / Deteriorating
Serviceability sewerage infrastructure	Assessment of the recent historical trend in serviceability to customers, as measured by movements in service and asset performance indicators.	Stable / Improving / Marginal /Deteriorating
Leakage	Total leakage measures the sum of distribution losses and supply pipe losses in megalitres per day (Ml/d). It includes any uncontrolled losses between the treatment works and the customer's stop tap. It does not include internal plumbing losses.	Megalitres a day (Ml/day)
Security of supply index (SoSI)	Indicates the extent to which a company is able to guarantee provision of its levels of service for restrictions of supply. The indicator measures security of supply for two scenarios (where relevant) – under dry year	Index score

Environmental impact

Indicator	Definition	Measure
Greenhouse gas (GHG) emissions	Measurement of the annual operational GHG emissions of the regulated business	Kilo tonnes of carbon dioxide equivalent (ktCO2e)
Pollution incidents (sewerage)	The total number of pollution incidents (categories 1 to 3) in a calendar year emanating from a discharge or escape of a contaminant from a sewerage company asset	Category 1-3 incidents per 10,000 km of sewer
Serious pollution incidents (sewerage)	The total number of serious pollution incidents (categories 1 and 2) in a calendar year emanating from a discharge or escape of a contaminant from a sewerage company asset.	Category 1-2 incidents per 10,000 km of sewer
Discharge permit compliance	Performance of sewerage assets to treat and dispose of sewage in line with the discharge permit conditions imposed on sewage treatment works.	%
Satisfactory sludge disposal	Companies determine their own definitions of satisfactory sludge disposal; as a minimum, we expect companies to adhere to the Safe Sludge Matrix and comply with any legal obligations.	%

Financial

Indicator	Definition	Measure
Post-tax return on capital	Current cost operating profit less tax as a return on regulatory capital value	%
Credit rating	The company's ability to comply with its licence requirement to maintain an investment grade credit rating	Assessment by rating agencies
Gearing	Traditionally financed companies – net debt as a percentage of the total regulatory capital value at the financial year end OR structured companies – as defined by company financial covenants	%
Interest cover	For traditionally financed companies, adjusted interest cover and FFO/interest OR for structured companies, adjusted interest cover or PMICR as required within the financial covenants. For reporting purposes, the company is required to report the lower of the interest cover ratios.	Ratio

Publishing Key Performance Indicators

We expect each company to publish all of the indicators that are relevant to the services they provide. Although we have set out the indicators that we expect companies to publish, we do not intend to produce a template for publication. It is for the companies to decide on a format that sets out this information transparently. We recognise that companies may wish to provide additional information to explain the measure or their performance.

The indicators include those that the Environment Agency will use to assess water company performance in order to protect the environment. Companies have told us that they also expect to publish indicators that cover the quality of their drinking water. They are responsible for determining the assurance processes that they need in order to provide accurate and reliable information.

Companies may choose how to present the indicators but they must be published with:

- status (red, amber or green, where indicated)
- unit of measurement as per the guidance
- actual value

Red: Not compliant with the guidance and having a material impact on reporting

Amber: Not compliant with the guidance and having no material impact on reporting.

Green: Fully-compliant with the guidance

Companies may choose to publish additional indicators that they have defined and that they consider are important as part of their engagement with their stakeholders. The customer challenge groups may also use the indicators to understand company performance.

We also recognise that for some indicators it would be useful to stakeholders to understand how performance compares with earlier years. But it is for the companies to decide how this is presented.

In presenting the indicators companies may wish to provide supplementary information on the size of their operations in order for stakeholders to draw comparisons across the water and sewerage sectors.

Customer Experience

Service Incentive Mechanism (SIM) Score

Repeat flooding incidents (Internal Sewer flooding) Number of incidents

Water supply interruptions Minute: Second per total properties served

Reliability and availability

Serviceability water non-infrastructure Stable/Improving/Marginal/Deteriorating

Serviceability water infrastructure Stable/Improving/Marginal/Deteriorating

Serviceability sewerage non-infrastructure Stable/Improving/Marginal/Deteriorating

Serviceability sewerage infrastructure Stable/Improving/Marginal/Deteriorating

Leakage Ml/day

Security of Supply Index (SOSI) Index score

Drinking Water Quality %

Environmental Impact

Greenhouse gas (GHG) emissions ktCO2e

Pollution incidents (sewerage) Category 1-3 incidents per 10,000 km of sewer

Serious pollution incidents (sewerage) Category 1-2 incidents per 10,000 km of sewer

Discharge permit compliance %

Satisfactory sludge disposal %

Financial

Post-tax return on capital %

Credit rating

Gearing %

Interest cover

The Guaranteed Standards Scheme (GSS)

There are also seven key issues where Ofwat set down (in April 2017) measures for reimbursement to customers who are adversely affected by poor service. We have resisted the temptation to bore you with the detail especially if you are resistant to bullet lists which contain all of the (excellent) reasons why a payment may not be made. You can find the detail on-line by Googling 'The Guaranteed Standards Scheme.'

1. Appointments

Making appointments

Keeping appointments

2. Complaints, account queries and requests about payment arrangements

Account queries and requests about changes to payment arrangements

Written complaints

3. Notice of interruption to supply

4. Supply not restored

5. Low pressure

A company must maintain a minimum pressure in the communication pipe of seven metres static head (0.7 bar).

6. Flooding from sewers – internal flooding

7. Flooding from sewers – external flooding

A detailed schedule of the payment requirements follows.

In-period ODI determinations

In 2019 Ofwat set water companies service targets ('performance commitments') for the period 2020-25. This would encourage companies to deliver on the objectives that matter to customers and the environment. They have rewards and penalties ('outcome delivery incentives (ODIs)') associated with them most of which are financial:

- If a company performs better than the performance commitment level, it may get a financial reward ('outperformance payments')

- If a company performs worse below the performance commitment level it may pay a financial penalty ('underperformance payments')

Each company is encouraged to deliver what customers want as they get outperformance payments for doing well and are penalised if they do badly.

[It's no wonder that two of the major water companies could not find time to assist with this publication!]

Appendix 7: Further Reading

[No list like this is ever complete]

The Design of Sewers and Sewage Treatment Works by JB White (note – JB, not EB)

Wastewater Engineering by JB White

Flushed with Pride by Wallace Reyburn

Taming the Flood by Jeremy Purseglove*

Working with Nature by Jeremy Purseglove

GB Water Industry for Dummies by Dr Graham Haimsworth

About Water Treatment by Kemira

Water Systems – Sky to Sea by Louise Halstrap at The Centre for Alternative Technology (Webinar)

Guide 27 – Ventilated Improved Pit latrines by WEDC at Loughborough University

The Water Book by Alok Jha

The Drinking Water Book by Colin Ingram

Water is Water by Miranda Paul and Jason Chin

Water by Melissa Stewart (childrens' book)

An Underground Guide to Sewers : or: Down, Through and Out in Paris, London, New York, etc. by Stephen Halliday

The Great Stink of London : Sir Joseph Bazalgette and the Cleansing of the Victorian Metropolis by Stephen Halliday

The Last Taboo : Opening the Door on the Global Sanitation Crisis published by UNICEF

Vertical Flow Constructed Wetlands by Stefanakis, Akratos and Tsihrintzis

Wastewater Treatment : Advanced Processes and Technologies edited by: Rao, Senthilkumar, Byrne, and Feroz

Wastewater Sludge Processing by Turovskiy and Mathai

Water and Waste: Four Hundred Years of Health Improvements in the Lea Valley by Jim Lewis

SAMS, Simplified Asset Management Systems by Pyotr Stilovsky

Principles of Asset Management by Styles and Earp

Twort's Water Supply Illustrated by Malcolm J Brandt, K Michael Johnson, Andrew J Elphinston and Don D Ratnayaka

From WRc:

- *Sewers for Adoption 7th Edition*
- *Civil Engineering Specification for the Water Industry 7th Edition*
- *Sewerage Risk Management*
- *Model Contract Documents*
- *Manual of Sewer Condition Classification*
- *Drain Repair Book 4th Edition*
- *Water Mains Rehabilitation Manual*

Confined spaces

A brief guide to working safely by HSE published 2013

* Jeremy Purseglove worked for Severn-Trent Water Authority in the 1980s when the MAFF philosophy was still in vogue in respect of land drainage and rivers. He argued vociferously to stop destroying our watercourses and for a return to nature, especially as a means of dealing with flooding. His message fell on deaf ears and the Director of Operations gave him a year's sabbatical to write his book – *Taming the Flood* – and 'found him another job'. He spent many years with Mott McDonald as their environmental advisor. He has since been a visiting professor and become chair of the group who run Wicken Fen.

The book is somewhat long and over detailed but the messages contained within are years ahead of their time.

Appendix 8: A Water Grid for England

This 'paper' is based on a submission to ICE following a consultation on the future of water resources.

Most water companies have an internal 'strategic grid' which enables the major water treatment plants to support eachother when there are problems. However, these grids are largely parochial in nature and, since the demise of the Water Resources Board, joined-up thinking has been in short supply. Once mention is made of a 'water grid' everyone and his dog come out to say that it's impossible and too expensive. Actually a substantive grid, moving raw water around between existing major reservoirs is both practical and economic – the likely overall cost being less than is being spent on the Thames Tideway project.

The key issues are:

- Only transfer raw water and
- Make optimal use of existing infrastructure

Overlaid on a diagram of the existing sources and assets, the proposed links (in red) shown below could form the basis of an affordable water grid which would solve much, though not all of the problem. The key components involve the interlinking of the North Midlands systems with the reservoirs which serve much of the Anglian region. Taking water from the Trent is dependent upon quality issues though these are not insurmountable especially when the choice is between providing expensive treatment or running out of water to treat. Transfers into the Thames catchment can be made through an extension of this system, from the Warwickshire Avon, from the Bristol Avon, from the Severn and even from the lower reaches of the Wye.

The areas south of the Thames and Kent are shown in the second diagram. This requires more lateral thinking than simply interlinking the raw water reservoirs. Kent and the South are inherently in deficit and the situation is not helped by the plethora of small private water companies.

Interlinking of the Kent reservoirs along similar principles to those suggested in the Midlands is a simple matter of geometry but the provision of additional resources is not. Importing water by tanker from the North East (Kielder) can be achieved by unloading at Gravesend to Bewl Water and at Southampton for Testwood Lakes. In addition, it should be possible to provide additional treatment for effluent from Crossness and transfer this for blending in Bewl Water.

Missing from the diagram are the systems bringing water from the Lake District to Manchester, and from Kielder Water to support the rivers south of there. Also Ladybower serves both Severn Trent and Sheffield under an agreement which runs out in 2030.

It took over ten years to interlink the East and West Midlands following the 1976 drought and twenty years for Severn Trent to complete its strategic grid. How long will this take?

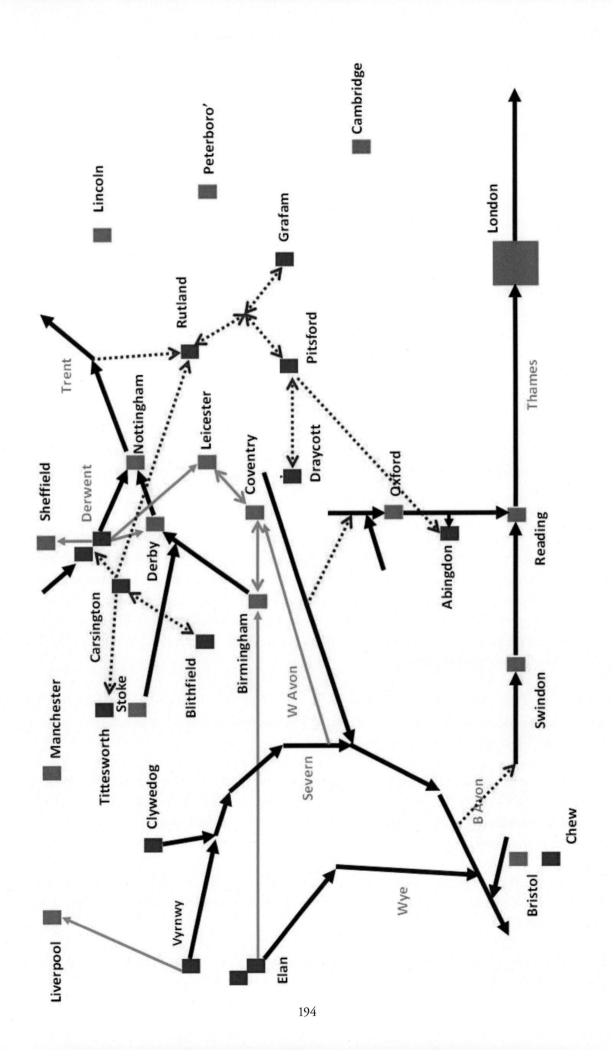

194

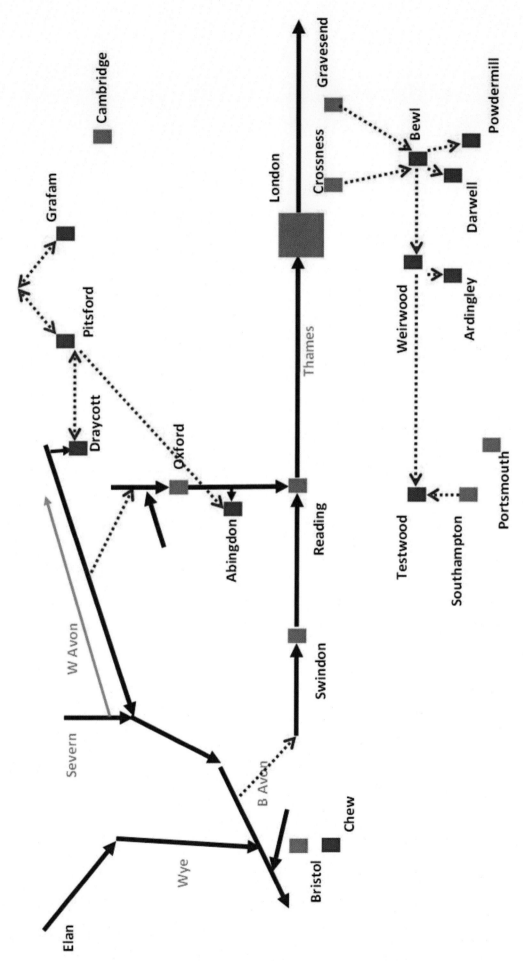

The river Rhine, it is well known,
Doth wash your city of Cologne;
But tell me, Nymphs, what power divine
Shall henceforth wash the river Rhine?

Samuel Taylor Coleridge, "Cologne" (1828).

Printed in the United States
by Baker & Taylor Publisher Services